STATISTICAL DISTRIBUTIONS

A HANDBOOK FOR STUDENTS AND PRACTITIONERS

STATISTICAL DISTRIBUTIONS

A HANDBOOK FOR STUDENTS AND PRACTITIONERS

N. A. J. Hastings
and
J. B. Peacock

A HALSTED PRESS BOOK

JOHN WILEY & SONS
New York—Toronto

English edition first published in 1975 by Butterworth & Co (Publishers) Ltd
88 Kingsway, London WC2B 6AB

Published in the U.S.A. and Canada by Halsted Press, a Division of John Wiley & Sons, Inc.,
New York

Library of Congress Cataloging in Publication Data

Hastings, N A J 1937-
Statistical distributions.

"A Halsted Press book."
Bibliography: p.
1. Distribution (Probability theory) I. Peacock,
J. B., joint author. II. Title.
QA273.6.H37 519.5'3 74-617

ISBN 0-470-35889-0

Text set in 10/12 pt IBM Press Roman, printed by photolithography, and bound in Great Britain at The Pitman Press, Bath

CONTENTS

List of Tables xi
List of Figures xi

1 PREFACE 1

2 TERMS AND SYMBOLS 2

2.1 Random variable, variate, random number 2
2.1.1 Probabilistic experiment 2
2.1.2 Possibility space 2
2.1.3 Random variable 3
2.1.4 Variate 3
2.1.5 Random number 3

2.2 Range, fractile, distribution function 3
2.2.1 Range 3
2.2.2 Fractile 4
2.2.3 Probability statement 4
2.2.4 Probability domain 4
2.2.5 Distribution function 5

2.3 Inverse distribution function 7
2.3.1 Inverse survival function 7

2.4 Probability density function and probability function 9

2.5 Other associated functions and quantities 9

3 GENERAL VARIATE RELATIONSHIPS 15

3.1 Introduction 15

3.2 Function of a variate 15

3.3 One to one transformations 16
3.3.1 Many to one function 16
3.3.2 Inverse of a one to one function 16

3.4 Variate relationships under one to one transformation 18
3.4.1 Probability statements 18
3.4.2 Distribution function 18
3.4.3 Inverse distribution function 18
3.4.4 Equivalence of variates 19
3.4.5 Inverse function of a variate 19

3.5 Parameters 20
3.5.1 Variate and function notation 20

3.6 Transformation of location and scale 21

3.7 Transformation from the rectangular variate 22

3.8 Many to one transformations 23
3.8.1 Example 23
3.8.2 Symmetrical distributions 23

3.9 Functions of several variates 25

4 BERNOULLI DISTRIBUTION 28

4.1 Random number generation 28

4.2 Curtailed Bernoulli trial sequences 28

5 BETA DISTRIBUTION 30

5.1 Parameter estimation 32

5.2 Notes on beta and gamma functions 32

5.3 Variate relationships 34

5.4 Random number generation 35

6 BINOMIAL DISTRIBUTION 36

6.1 Parameter estimation 38

6.2 Variate relationships 38

6.3 Note 40

6.4 Random number generation 41

7 CAUCHY DISTRIBUTION 42

7.1 Variate relationships 42

7.2 Random number generation 44

7.3 General form 45

7.4 Notes 45

8 CHI-SQUARED DISTRIBUTION 46

8.1 Variate relationships 46

8.2 Random number generation 51

9 DISCRETE UNIFORM DISTRIBUTION 52

9.1 Parameter estimation 52

10 ERLANG DISTRIBUTION 54

10.1 Parameter estimation 55

10.2 Variate relationships 55

10.3 Random number generation 55

11 EXPONENTIAL DISTRIBUTION 56

11.1 Parameter estimation 58

11.2 Variate relationships 58

11.3 Random number generation 58

12 EXTREME VALUE DISTRIBUTION 60

12.1 Variate relationships 62

12.2 Random number generation 62

13 F DISTRIBUTION (Variance ratio distribution) 64

13.1 Variate relationships 64

14 GAMMA DISTRIBUTION 68

14.1 Parameter estimation 70

14.2 Variate relationships 70

14.3 Random number generation 72

15 GEOMETRIC DISTRIBUTION 74

15.1 Parameter estimation 77

15.2 Variate relationships 77

15.3 Random number generation 77

16 HYPERGEOMETRIC DISTRIBUTION 78

16.1 Parameter estimation 78

16.2 Variate relationship 79

16.3 Random number generation 79

16.4 Note 79

17 LOGISTIC DISTRIBUTION 80

17.1 Random number generation 82

18 LOGNORMAL DISTRIBUTION 84

18.1 Parameter estimation 86

18.2 Variate relationships 86

18.3 Random number generation 88

19 MULTINOMIAL DISTRIBUTION 90

20 NEGATIVE BINOMIAL DISTRIBUTION 92

20.1 Parameter estimation 94

20.2 Variate relationships 94

20.3 Notes 94

20.4 Random number generation 95

21 NORMAL DISTRIBUTION 96

21.1 Parameter estimation 98

21.2 Variate relationships 98

21.3 Random number generation 100

22 PARETO DISTRIBUTION 102

22.1 Parameter estimation 102

22.2 Variate relationship 104

22.3 Random number generation 104

23 PASCAL DISTRIBUTION 106

23.1 Parameter estimation 106

23.2 Variate relationships 106

23.3 Random number generation 107

24 POISSON DISTRIBUTION 108

24.1 Parameter estimation 110

24.2 Variate relationships 110

24.3 Random number generation 112

24.4 Note 112

25 POWER FUNCTION DISTRIBUTION 114

26 RECTANGULAR DISTRIBUTION (Continuous uniform distribution) 116

26.1 Parameter estimation 118

26.2 Variate relationships 118

27 STUDENT'S T DISTRIBUTION 120

27.1 Variate relationships 120

28 WEIBULL DISTRIBUTION 124

28.1 Parameter estimation 126

28.2 Variate relationships 128

28.3 Random number generation 128

28.4 Note 128

BIBLIOGRAPHY 130

TABLES

Table 2.1 Functions and related quantities for a general variate 10

Table 2.2 General relationships between moments 14

Table 2.3 Samples 14

Table 3.1 Relationships between functions for variates which differ only by location and scale parameters a, b 22

FIGURES

Figure 2.1 The random variable 'number of heads' 2

Figure 2.2 The distribution function F: $x \to \alpha$ or $\alpha = F(x)$ for the random variable 'number of heads' 5

Figure 2.3 Distribution function and inverse distribution function for a continuous variate 6

Figure 2.4 Distribution function and inverse distribution function for a discrete variate 6

Figure 2.5 Probability density function 8

Figure 2.6 Probability density function: illustrating the fractile corresponding to a given probability α. G is the inverse distribution function 8

Figure 3.1 A one to one function 17

Figure 3.2 A many to one function 17

Figure 5.1 Probability density function for the beta variate $\boldsymbol{\beta}$: ν, ω 31

Figure 5.2 Probability density function for the beta variate $\boldsymbol{\beta}$: ν, ω for additional values of the parameters 31

Figure 5.3 Distribution function for the beta variate $\boldsymbol{\beta}$: ν, ω 33

Figure 6.1 Probability function for the binomial variate **B**: n,p 37

Figure 6.2 Distribution function for the binomial variate **B**: n,p 39

Figure 7.1 Cauchy probability density function 43

Figure 8.1 Probability density function for the chi-squared variate $\boldsymbol{\chi}^2$: ν 47

Figure 8.2 Distribution function for the chi-squared variate $\boldsymbol{\chi}^2$: ν 49

Figure 9.1 Probability density function for the discrete uniform variate **D**: a, b 53

Figure 9.2 Distribution function for the discrete uniform variate **D**: a, b 53

Figure 11.1 Probability density function for the exponential variate **E**: b 57

Figure 11.2 Distribution function for the exponential variate **E**: b 57

Figure 11.3 Cumulative hazard function for the exponential variate **E**: b 57

Figure 11.4 Exponential distribution, probability paper 59

Figure 12.1 Probability density function for the extreme value variate **X**: a, b (smallest extreme) 61

Figure 12.2 Distribution function for the extreme value variate **X**: a, b (smallest extreme) 61

Figure 12.3 Hazard function for the extreme value variate **X**: a, b (smallest extreme) 63

Figure 13.1 Probability density function for the **F** variate **F**: 4, 40 65

Figure 13.2 Distribution function for the **F** variate **F**: 4, 40 65

Figure 14.1 Probability density function for the gamma variate **γ** : 1, c 69

Figure 14.2 Distribution function for the gamma variate **γ** : 1, c 71

Figure 14.3 Hazard function for the gamma variate **γ** : 1, c 73

Figure 15.1 Probability function for the geometric variate **G**: p 75

Figure 17.1 Probability density function for the logistic variate with $a = 0, k = 1$ 81

Figure 17.2 Distribution function for the logistic variate with $a = 0, k = 1$ 83

Figure 18.1 Probability density function for the lognormal variate **L**: m, σ 85

Figure 18.2 Distribution function for the lognormal variate **L**: m, σ 85

Figure 18.3 Hazard function for the lognormal variate **L**: m, σ 87

Figure 18.4 Lognormal distribution, probability paper 89

Figure 20.1 Probability function for the negative binomial variate **Y**: x, p 93

Figure 21.1 Probability density function for the standard normal variate **N**: 0, 1 97

Figure 21.2 Distribution function for the standard normal variate **N**: 0, 1 97

Figure 21.3 Hazard function for the standard normal variate **N**: 0, 1 99

Figure 21.4 Normal distribution, probability paper 101

Figure 22.1 Probability density function for the Pareto variate 103

Figure 22.2 Distribution function for the Pareto variate 105

Figure 24.1 Probability function for the Poisson variate **P**: λ 109

Figure 24.2 Distribution function for the Poisson variate **P**: λ 111

Figure 24.3 Curves for the Poisson variate **P**: λ, showing p = Prob [P: $\lambda \geqslant x$] 113

Figure 25.1 Probability density function for the power function variate 115

Figure 25.2 Distribution function for the power function variate 115

Figure 26.1 Probability density function for the rectangular variate **R**: a, b 117

Figure 26.2 Distribution function for the rectangular variate **R**: a, b 117

Figure 26.3 Hazard function for the unit rectangular variate **R**: 0, 1 119

Figure 27.1 Probability density function for Student's **T** variate with one degree of freedom 121

Figure 27.2 Distribution function for Student's **T** variate with one degree of freedom 121

Figure 28.1 Probability density function for the Weibull variate **W**: 1, c 125

Figure 28.2 Distribution function for the Weibull variate **W**: 1, c 125

Figure 28.3 Hazard function for the Weibull variate **W**: 1, c 127

Figure 28.4 Weibull mean and standard deviation (S.D.) as a function of the shape parameter c and scale parameter b 127

Figure 28.5 Weibull probability chart 129

1 PREFACE

The number of puppies in a litter, the life of a light bulb, the time to the arrival of the next bus at a stop; these are all examples of random variables which are encountered in everyday life. Random variables have come to play an important role in nearly every field of study; in physics, chemistry and engineering and especially in the biological, social and management sciences. Random variables are measured and analysed in terms of their statistical and probabilistic properties, an underlying feature of which is the distribution function. Although the number of potential distribution models is very large, in practice a relatively small number have come to prominence, either because they have desirable mathematical characteristics or because they relate particularly well to some slice of reality or both.

This book gives a concise statement of leading facts relating to 25 distributions, and includes diagrams so that shapes and other general properties may be readily appreciated. Probability papers are included in appropriate cases. A consistent system of nomenclature is used throughout. We have found ourselves in need of just such a summary on frequent occasions; as students, as teachers and as practitioners. This book has been prepared in an attempt to fill the need for rapid access to information which must otherwise be gleaned from scattered and individually costly sources.

In choosing the material we have been guided by a utilitarian outlook, so that for example some distributions which are special cases of more general families are given separate treatment where this is felt to be justified by applications. In the choosing of appropriate symbols and parameters for the description of each distribution, and especially where different but interrelated sets of symbols are in use in different fields, we have tried to strike a balance between the various usages, the need for a consistent system of nomenclature within the book and typographic simplicity.

In addition to listing the properties of individual variates we have considered relationships between variates. This area is often obscure to the non-specialist. We have also made use of the inverse distribution function, a function which is widely tabulated and used but rarely explicitly defined. We have particularly sought to avoid the confusion which can result from using a single symbol (a notorious example is χ^2) to mean here a function, there a fractile and elsewhere a variate.

2 TERMS AND SYMBOLS

2.1 Random variable, variate, random number

2.1.1 Probabilistic experiment

A probabilistic experiment is some occurrence such as the tossing of coins, rolling dice or observation of rainfall on a particular day, where a complex natural background leads to a chance outcome.

2.1.2 Possibility space

The set of possible outcomes of a probabilistic experiment is called a possibility space. For example, if two coins are tossed the possibility space is the set of possible results HH, HT, TH, TT, where H indicates a head and T a tail.

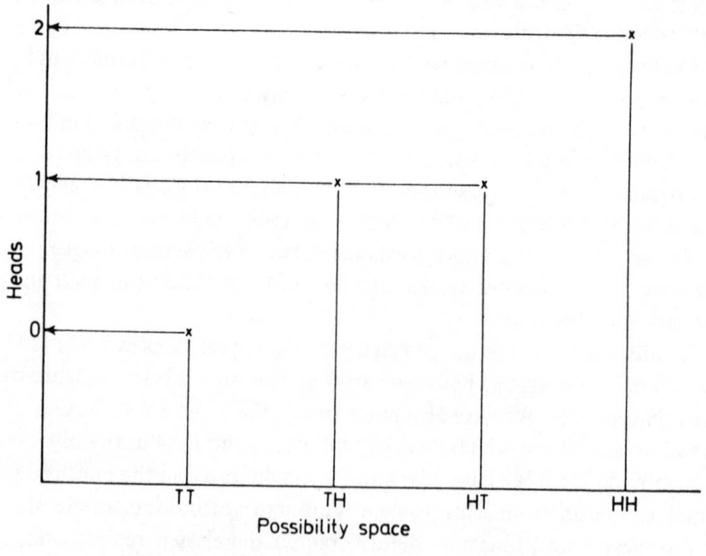

Figure 2.1 The random variable 'number of heads'

2.1.3 *Random variable*

A random variable is a function which maps from a possibility space into a set of numbers of something. Several different random variables may be defined in relation to a given experiment. Thus in the case of tossing two coins the number of heads observed is one random variable, the number of tails is another and the number of double heads is another. The random variable 'number of heads' associates the number zero with the event TT, the number 1 with the events TH and HT and the number 2 with the event HH. *Figure 2.1* illustrates this mapping.

2.1.4 *Variate*

In the discussion of statistical distributions it is convenient to work in terms of variates. A variate is a generalisation of the idea of a random variable and has similar probabilistic properties but is defined without reference to a particular type of probabilistic experiment. A *variate* is the set of all random variables which obey a given probabilistic law. The number of heads and the number of tails observed in independent coin tossing experiments are elements of the same variate since the probabilistic factors governing the numerical part of their outcome are identical.

2.1.5 *Random number*

A *random number* associated with a given variate is a number generated at a realisation of any random variable which is an element of that variate.

2.2 Range, fractile, distribution function

2.2.1 *Range*

Let $\mathbf{X}$ denote a variate and let $\mathcal{R}_{\mathbf{X}}$ be the set of all (real number) values which the variate can take. $\mathcal{R}_{\mathbf{X}}$ is the *range* of $\mathbf{X}$. As an illustration (illustrations are in terms of random variables) consider the experiment of tossing two coins and noting the number of heads. The range of this random variable is the set $\{0, 1, 2\}$ heads, since the result may show zero, one or two heads.

2.2.2 *Fractile*

For a general variate **X** let x (a real number) denote a general element of the range $\mathcal{R}_{\mathbf{X}}$. We refer to x as the *fractile* of **X**. In the coin tossing experiment referred to in Section 2.2.1 $x \in \{0, 1, 2\}$ heads, that is, x is a member of the set $\{0, 1, 2\}$ heads.

2.2.3 *Probability statement*

Let $\mathbf{X} = x$ mean 'the value realised by the variate **X** is x'.
Let Prob $[\mathbf{X} \leqslant x]$ mean 'the probability that the value realised by the variate **X** is less than or equal to x'.

2.2.4 *Probability domain*

Let α (a real number) denote probability. Let $\mathcal{R}_{\mathbf{X}}^{\alpha}$ be the set of all values (of probability) that Prob $[\mathbf{X} \leqslant x]$ can take. For a continuous variate $\mathcal{R}_{\mathbf{X}}^{\alpha}$ is the line segment 0, 1; for a discrete variate it will be a subset of that segment. $\mathcal{R}_{\mathbf{X}}^{\alpha}$ is the *probability domain* of the variate **X**.

In examples we shall use the symbol **X** to denote a random variable. Let **X** be the number of heads observed when two coins are tossed. We then have

$$\text{Prob }[\mathbf{X} \leqslant 0] = \tfrac{1}{4}$$

$$\text{Prob }[\mathbf{X} \leqslant 1] = \tfrac{3}{4}$$

$$\text{Prob }[\mathbf{X} \leqslant 2] = 1$$

and hence

$$\mathcal{R}_{\mathbf{X}}^{\alpha} = \left\{\tfrac{1}{4}, \tfrac{3}{4}, 1\right\}$$

2.2.5 Distribution function

The *distribution function* F (or more specifically $F_{\mathbf{X}}$) associated with a variate $\mathbf{X}$ is a function which maps from the range $\mathcal{R}_{\mathbf{X}}$ into the probability domain $\mathcal{R}_{\mathbf{X}}^{\alpha}$ and which is such that

$$F(x) = \text{Prob}\ [\mathbf{X} \leqslant x] = \alpha \qquad x \in \mathcal{R}_{\mathbf{X}},\ \alpha \in \mathcal{R}_{\mathbf{X}}^{\alpha} \tag{2.2a}$$

$F(x)$ is increasing in x and attains the value unity at the maximum of x. *Figure 2.2* illustrates the distribution function for the number of heads in the experiment of tossing two coins, *Figure 2.3* illustrates a general continuous distribution function and *Figure 2.4* a general discrete distribution function.

A distribution function performs a one to one mapping from $\mathcal{R}_{\mathbf{X}}$ into $\mathcal{R}_{\mathbf{X}}^{\alpha}$ and is therefore a one to one function (as discussed further in Section 3).

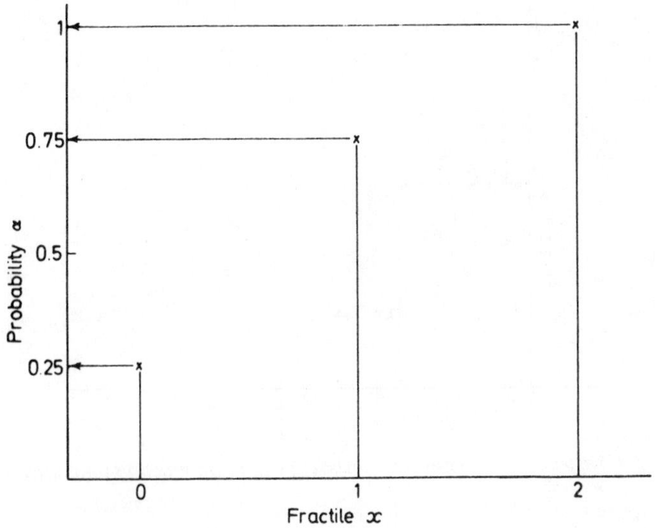

Figure 2.2 The distribution function F: x → α or α = F(x) for the random variable 'number of heads'

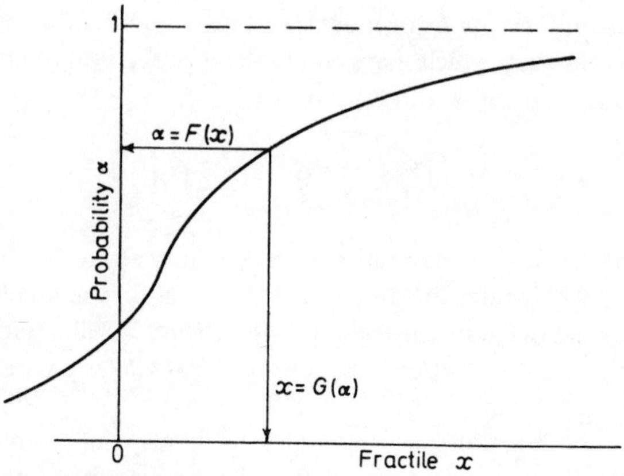

Figure 2.3 Distribution function and inverse distribution function for a continuous variate

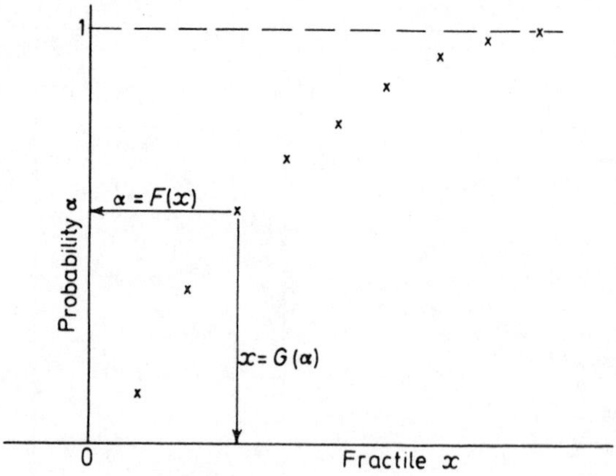

Figure 2.4 Distribution function and inverse distribution function for a discrete variate

2.3 Inverse distribution function

For a distribution function F, mapping a fractile x into a probability α, the inverse distribution function G performs the corresponding inverse mapping from α into x. Thus for $x \in \mathcal{R}_{\mathbf{X}}$, $\alpha \in \mathcal{R}_{\mathbf{X}}^{\alpha}$, the following statements hold

$$\alpha = F(x) \tag{2.3a}$$

$$x = G(\alpha) \tag{2.3b}$$

$$x = G(F(x)) \tag{2.3c}$$

$$\alpha = F(G(\alpha)) \tag{2.3d}$$

$$\text{Prob}\ [\mathbf{X} \leqslant x] = F(x) = \alpha \tag{2.3e}$$

$$\text{Prob}\ [\mathbf{X} \leqslant G(\alpha)] = F(x) = \alpha \tag{2.3f}$$

$G(\alpha)$ is the fractile such that the probability that the variate takes a value less than or equal to it is α.

Figures 2.2, 2.3 and *2.4* illustrate both distribution functions and inverse distribution functions, the difference lying only in the choice of independent variable.

For the two-coin tossing experiment the distribution function, F, and inverse distribution function, G, of the number of heads are as follows

$$F(0) = \tfrac{1}{4} \qquad G(\tfrac{1}{4}) = 0$$

$$F(1) = \tfrac{3}{4} \qquad G(\tfrac{3}{4}) = 1$$

$$F(2) = 1 \qquad G(1) = 2$$

2.3.1 *Inverse survival function*

The inverse survival function Z is a function such that $Z(\alpha)$ is the fractile which is exceeded with probability α. This definition leads to the following equations

$$\text{Prob}\ [\mathbf{X} > Z(\alpha)] = \alpha \tag{2.3g}$$

$$Z(\alpha) = G(1 - \alpha) \tag{2.3h}$$

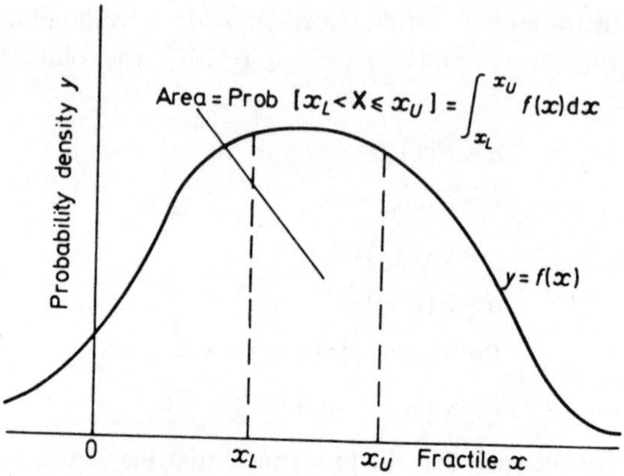

Figure 2.5 Probability density function

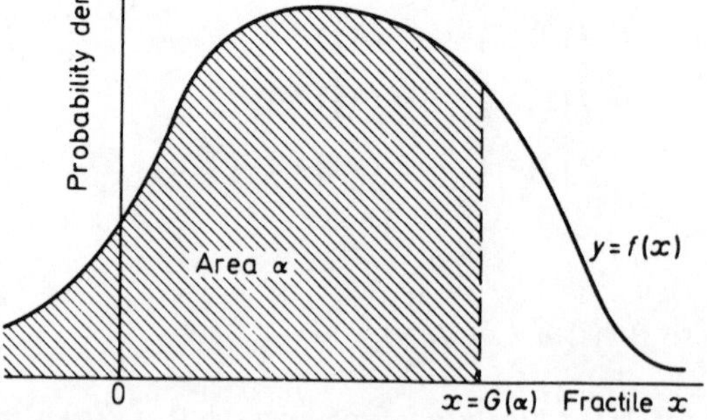

Figure 2.6 Probability density function: illustrating the fractile corresponding to a given probability α. G is the inverse distribution function

Inverse survival functions are among the most widely tabulated functions in statistics. For example, the well known chi-squared tables are tables of the fractile x as a function of the probability level α and a shape parameter, and are tables of the chi-squared inverse survival function.

2.4 Probability density function and probability function

A probability density function, $f(x)$, is the first differential coefficient of a distribution function, $F(x)$, with respect to x (where this differential exists).

$$f(x) = \frac{\mathrm{d}(F(x))}{\mathrm{d}x} \tag{2.4a}$$

For a given continuous variate **X** the area under the probability density curve between two points x_L, x_U in the range of **X** is equal to the probability that an as yet unrealised random number of **X** will lie between x_L and x_U. *Figure 2.5* illustrates this. *Figure 2.6* illustrates the relationship between the area under a probability density curve and the fractile mapped by the inverse distribution function at the corresponding probability value.

A discrete variate takes discrete values x with finite probabilities $f(x)$. In this case $f(x)$ is the probability function, also called the probability mass function.

2.5 Other associated functions and quantities

In addition to the functions just described there are many other functions and quantities which are associated with a given variate. A listing is given in *Table 2.1* relating to a general variate which may be either continuous or discrete. The integrals in *Table 2.1* are Stieltjes integrals which for discrete variates become ordinary summations, so

$$\int_{x_L}^{x_U} \phi(x) f(x)\,\mathrm{d}x \text{ corresponds to } \sum_{x=x_L}^{x_U} \phi(x) f(x) \tag{2.5a}$$

Table 2.2 gives some general relationships between moments and *Table 2.3* gives our notation for values, mean and variance for samples.

Terms and symbols

Table 2.1 FUNCTIONS AND RELATED QUANTITIES FOR A GENERAL VARIATE

X denotes a variate, x denotes a fractile and α denotes probability.

Term	*Symbol*	*Description and Notes*
1. Distribution function (d.f.), or cumulative distribution function (c.d.f.)	$F(x)$	$F(x)$ is the probability that the variate takes a value less than or equal to x. $F(x) = \text{Prob}\,[\mathbf{X} \leqslant x] = \alpha$; $F(x) = \int_{-\infty}^{x} f(u)\,\mathrm{d}u$
2. Probability density function (p.d.f.)	$f(x)$	A function whose integral over the range x_L to x_U is equal to the probability that the variate takes a value in that range. $\int_{x_L}^{x_U} f(x)\,\mathrm{d}x = \text{Prob}\,[x_L < \mathbf{X} \leqslant x_U]$; $f(x) = \dfrac{\mathrm{d}(F(x))}{\mathrm{d}x}$
3. Probability function (discrete variates)	$f(x)$	$f(x)$ is the probability that the variate takes the value x. $f(x) = \text{Prob}\,[\mathbf{X} = x]$.
4. Inverse distribution function (of probability α)	$G(\alpha)$	$G(\alpha)$ is the fractile such that the probability that the variate takes a value less than or equal to it is α. $x = G(\alpha) = G(F(x))$; $\text{Prob}\,[\mathbf{X} \leqslant G(\alpha)] = \alpha$. $G(\alpha)$ is the 100α percentage point. The relation to d.f. and p.d.f. is shown in *Figures 2.3, 2.4* and *2.6*.
5. Survival function	$S(x)$	$S(x)$ is the probability that the variate takes a value greater than x. $S(x) = \text{Prob}\,[\mathbf{X} > x] = 1 - F(x)$
6. Inverse survival function (of probability α)	$Z(\alpha)$	$Z(\alpha)$ is the fractile which is exceeded with probability α. $\text{Prob}\,[\mathbf{X} > Z(\alpha)] = \alpha$. $x = Z(\alpha) = Z(S(x))$ where S is the survival function. $Z(\alpha) = G(1-\alpha)$ where G is the inverse distribution function.

Table 2.1 FUNCTIONS AND RELATED QUANTITIES FOR A GENERAL VARIATE (*continued*)

Term	*Symbol*	*Description and Notes*
7. Hazard function (force of mortality)	$h(x)$	$h(x)$ is the ratio of the probability density to the survival function at fractile x. $h(x) = f(x)/S(x) = f(x)/(1 - F(x))$
8. Hazard function (discrete variates)	$h(x)$	$h(x) = f(x + 1)/(1 - F(x))$
9. Mills ratio	$m(x)$	$m(x) = (1 - F(x))/f(x) = 1/h(x)$
10. Cumulative hazard function	$H(x)$	Integral of the hazard function. $H(x) = \int_{-\infty}^{x} h(u)\,\mathrm{d}u$ $H(x) = -\log(1 - F(x))$ $S(x) = 1 - F(x) = \exp(-H(x))$
11. Probability generating function (discrete variates). Also called the geometric or z-transform	$P(t)$	A function of an auxiliary variable t (or z) such that the coefficient of $t^x = f(x)$. $P(t) = \sum_{x=0}^{\infty} t^x f(x),\ \mathbf{X} > 0$ $f(x) = (1/x!)\left(\frac{\partial^x P(t)}{\partial t^x}\right)_{t=0}$
12. Moment generating function (m.g.f.)	$M(t)$	A function of an auxiliary variable t whose general term is of the form $\mu_r' t^r/r!$ $M(t) = \int_{-\infty}^{+\infty} \exp(tx) f(x)\,\mathrm{d}x$ $M(t) = 1 + \mu_1' t + \mu_2' t^2/2! + \ldots + \mu_r' t^r/r! + \ldots$ For any variates $\mathbf{A}$ and $\mathbf{B}$ whose moment generating functions, $M_{\mathbf{A}}(t)$ and $M_{\mathbf{B}}(t)$ exist, $M_{\mathbf{A}+\mathbf{B}}(t) = M_{\mathbf{A}}(t) . M_{\mathbf{B}}(t)$
13. Laplace transform of the p.d.f.	$f^*(s)$	A function of the auxiliary variable s defined by $f^*(s) = \int_{0}^{\infty} \exp(-sx) f(x)\,\mathrm{d}x,\ \mathbf{X} \geqslant 0$

Table 2.1 FUNCTIONS AND RELATED QUANTITIES FOR A GENERAL VARIATE (*continued*)

Term	*Symbol*	*Description and Notes*
14. Characteristic function	$C(t)$	A function of the auxiliary variable t and the imaginary quantity i ($i^2 = -1$) which exists and is unique to a given p.d.f. $C(t) = \int_{-\infty}^{+\infty} \exp(itx) f(x)\,dx$ If $C(t)$ be expanded in powers of t and if μ'_r exists the general term is $\mu'_r (it)^r/r!$ For any variates **A** and **B**, $C_{\mathbf{A}+\mathbf{B}}(t) = C_{\mathbf{A}}(t) . C_{\mathbf{B}}(t)$
15. Cumulant function	$K(t)$	$K(t) = \log C(t)$ $K_{\mathbf{A}+\mathbf{B}}(t) = K_{\mathbf{A}}(t) + K_{\mathbf{B}}(t)$
16. r^{th} cumulant	k_r	The coefficient of $(it)^r/r!$ in the expansion of $K(t)$.
17. r^{th} moment about the origin	μ'_r	$\mu'_r = \int_{-\infty}^{+\infty} x^r f(x)\,dx$ $\mu'_r = \left(\frac{\partial^r M(t)}{\partial t^r}\right)_{t=0} = (-i)^r \left(\frac{\partial^r C(t)}{\partial t^r}\right)_{t=0}$
18. Mean (first moment about the origin)	μ	$\mu = \int_{-\infty}^{+\infty} x f(x)\,dx = \mu'_1$
19. r^{th} moment about the mean	μ_r	$\mu_r = \int_{-\infty}^{+\infty} (x-\mu)^r f(x)\,dx$
20. Variance (second moment about the mean)	σ^2	$\sigma^2 = \int_{-\infty}^{+\infty} (x-\mu)^2 f(x)\,dx = \mu_2$
21. Standard deviation (S.D.)	σ	The positive square root of the variance.

Table 2.1 FUNCTIONS AND RELATED QUANTITIES FOR A GENERAL VARIATE (*continued*)

Term	*Symbol*	*Description and Notes*
22. Mean deviation		$\int_{-\infty}^{+\infty} \lvert x-\mu\rvert f(x)\,dx$. The mean absolute value of the deviation from the mean.
23. Mode		A fractile for which the p.d.f. is a local maximum.
24. Median	m	The fractile which is exceeded with probability $\frac{1}{2}$. $m = G(\frac{1}{2})$.
25. Standardised r^{th} moment about the mean	η_r	The rth moment about the mean scaled so that the standard deviation is unity. $\eta_r = \int_{-\infty}^{+\infty} \left(\frac{x-\mu}{\sigma}\right)^r f(x)\,dx = \mu_r/\sigma^r$
26. Coefficient of skewness	η_3	$\eta_3 = \mu_3/\sigma^3$
27. Coefficient of kurtosis	η_4	$\eta_4 = \mu_4/\sigma^4$
28. Coefficient of variation		(S.D.)/Mean = σ/μ
29. Information content	I	$I = -\int_{-\infty}^{+\infty} f(x)\log_2(f(x))\,dx$
30. r^{th} factorial moment about the origin (discrete variates)	$\mu'_{(r)}$	$\sum_{x=0}^{\infty} f(x)\,.\,x(x-1)(x-2)\ldots(x-r+1)$, $\mathbf{X} \geqslant 0$ $\mu'_{(r)} = \left(\frac{\partial^r P(t)}{\partial t_r}\right)_{t=1}$
31. r^{th} factorial moment about the mean (discrete variate)	$\mu_{(r)}$	$\sum_{x=0}^{\infty} f(x-\mu)\,.\,(x-\mu)(x-\mu-1)\ldots(x-\mu-r+1)$, $\mathbf{X} \geqslant 0$

Table 2.2 GENERAL RELATIONSHIPS BETWEEN MOMENTS

$$\mu'_r = \sum_{i=0}^{r} \binom{r}{i} \mu_{r-i} (\mu'_1)^i$$

$$\mu_r = \sum_{i=0}^{r} \binom{r}{i} \mu'_{r-i} (-\mu'_1)^i$$

$\mu_0 = \mu'_0 = 1, \quad \mu_1 = 0$

Table 2.3 SAMPLES

Term	*Symbol*	*Description and Notes*
Sample data	x_i	x_i is an observed value of a random variable
Sample size	n	The number of observations in a sample
Sample mean	$\bar{x}$	$(1/n) \sum_{i=1}^{n} x_i$
Sample variance (unadjusted)	s^2	$(1/n) \sum_{i=1}^{n} (x_i - \bar{x})^2$

3 GENERAL VARIATE RELATIONSHIPS

3.1 Introduction

This section is concerned with general relationships between variates and with the ideas and notation needed to describe them. Some definitions are given and the relationships between variates under one to one transformation are developed. Location, scale and shape parameters are then introduced and the relationships between functions associated with variates which differ only in regard to location and scale are listed. The relationship of a general variate to the rectangular variate is derived and finally the notation and concepts involved in dealing with variates which are related by many to one functions and by functionals are discussed.

Following the notation introduced in Section 2 we denote a general variate by $\mathbf{X}$, its range by $\mathcal{R}_{\mathbf{X}}$, its fractile by x and a random number of $\mathbf{X}$ by $x_{\mathbf{X}}$.

3.2 Function of a variate

Let ϕ be a function mapping from $\mathcal{R}_{\mathbf{X}}$ into a set which we shall call $\mathcal{R}_{\phi(\mathbf{X})}$.

*Definition 3.2*a. *Function of a Variate:* $\phi(\mathbf{X})$ is a variate such that if $x_{\mathbf{X}}$ is a random number of $\mathbf{X}$ then $\phi(x_{\mathbf{X}})$ is a random number of $\phi(\mathbf{X})$.

Thus a function of a variate is itself a variate whose value at any realisation is obtained by applying the appropriate transformation to the value realised by the original variate. For example if $\mathbf{X}$ is the number of heads obtained when three coins are tossed then $\mathbf{X}^3$ is the cube of the number of heads obtained. (Here as in Section 2 we use the symbol $\mathbf{X}$ for both a variate and a random variable which is an element of that variate.)

The probabilistic relationship between $\mathbf{X}$ and $\phi(\mathbf{X})$ will depend on whether more than one number in $\mathcal{R}_{\mathbf{X}}$ maps into the same $\phi(x)$ in $\mathcal{R}_{\phi(\mathbf{X})}$. That is to say, it is important to consider whether ϕ is or is not a one to one function over the range considered. This point is taken up in Section 3.3.

A definition similar to 3.2a applies in the case of a function of several variates; we shall detail the case of a function of two variates. Let **X**, **Y** be variates with ranges $\mathcal{R}_\mathbf{X}$, $\mathcal{R}_\mathbf{Y}$ and let ψ be a functional mapping from the Cartesian product of $\mathcal{R}_\mathbf{X}$ and $\mathcal{R}_\mathbf{Y}$ into (all or part of) the real line.

Definition 3.2b. Function of Several Variates: $\psi(\mathbf{X}, \mathbf{Y})$ is a variate such that if $x_\mathbf{X}, x_\mathbf{Y}$ are random numbers of **X** and **Y** respectively then $\psi(x_\mathbf{X}, x_\mathbf{Y})$ is a random number of $\psi(\mathbf{X}, \mathbf{Y})$.

3.3 One to one transformations

Let ϕ be a function mapping from the real line into the real line.

Definition 3.3a. One to One Function: ϕ is a one to one function if there are no two numbers x_1, x_2 in the domain of ϕ such that $\phi(x_1) = \phi(x_2), x_1 \neq x_2$.

A sufficient condition for a function to be one to one is that it be increasing in x. As an example $\phi(x) = e^x$ is a one to one function but $\phi(x) = x^2$ is not (unless x is confined to all negative or all positive values, say) since $x_1 = 2$ and $x_2 = -2$ give $\phi(x_1) = \phi(x_2) = 4$. *Figures 3.1* and *3.2* illustrate this.

3.3.1 Many to one function

A function which is not one to one is many to one. See also Section 3.8.

3.3.2 Inverse of a one to one function

The inverse of a one to one function ϕ is a one to one function ϕ^{-1} where $\phi^{-1}(\phi(x)) = x$, $\phi(\phi^{-1}(y)) = y$ and x and y real numbers (Bernstein's theorem).

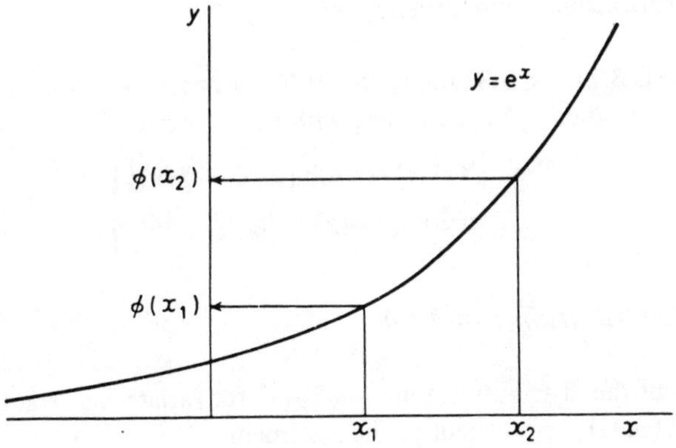

Figure 3.1 A one to one function

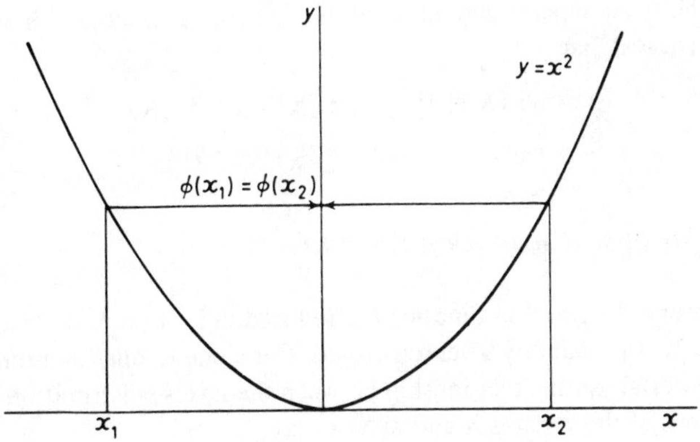

Figure 3.2 A many to one function

3.4 Variate relationships under one to one transformation

3.4.1 Probability statements

Definitions 3.2a and 3.3a imply that if **X** is a variate and ϕ is a one to one function then $\phi(\mathbf{X})$ is a variate with the property

$$\left.\begin{array}{c}\text{Prob}\,[\mathbf{X} \leqslant x] = \text{Prob}\,[\phi(\mathbf{X}) \leqslant \phi(x)] \\ x \in \mathcal{R}_{\mathbf{X}};\ \phi(x) \in \mathcal{R}_{\phi(\mathbf{X})}\end{array}\right\} \tag{3.4a}$$

3.4.2 Distribution function

In terms of the distribution function $F_{\mathbf{X}}(x)$ for variate **X** at fractile x, equation (3.4a) is equivalent to the statement

$$F_{\mathbf{X}}(x) = F_{\phi(\mathbf{X})}(\phi(x)) \tag{3.4b}$$

To illustrate equations (3.4a) and (3.4b) consider the experiment of tossing three coins and the random variables 'number of heads', denoted by **X** and 'cube of the number of heads', denoted by $\mathbf{X}^3$. The probability statements and distribution functions at fractiles 2 heads and 8 $(\text{heads})^3$ are

$$\left.\begin{array}{c}\text{Prob}\,[\mathbf{X} \leqslant 2] = \text{Prob}\,[\mathbf{X}^3 \leqslant 8] = 7/8 \\ F_{\mathbf{X}}(2) = F_{\mathbf{X}^3}(2^3) = F_{\mathbf{X}^3}(8) = 7/8\end{array}\right\} \tag{3.4c}$$

3.4.3 Inverse distribution function

The inverse distribution function (introduced in Section 2.3) for a variate **X** at probability level α is $G_{\mathbf{X}}(\alpha)$. For a one to one function ϕ we now establish the relationship between the inverse distribution functions of the variates **X** and $\phi(\mathbf{X})$.

*Theorem 3.4*a: $\phi(G_{\mathbf{X}}(\alpha)) = G_{\phi(\mathbf{X})}(\alpha)$

Proof: Equations (2.3f) and (3.4b) imply that if

$$G_{\mathbf{X}}(\alpha) = x \quad \text{then} \quad G_{\phi(\mathbf{X})}(\alpha) = \phi(x) \tag{3.4d}$$

Statement (3.4d) implies that Theorem 3.4a is true. QED

We illustrate Theorem 3.4a by extending the example of equation (3.4c). Considering the inverse distribution function we have

$$G_{\mathbf{X}}(7/8) = 2; G_{\mathbf{X}^3}(7/8) = 8 = 2^3 = (G_{\mathbf{X}}(7/8))^3 \qquad (3.4e)$$

3.4.4 *Equivalence of variates*

For any two variates **X** and **Y** the statement $\mathbf{X} \sim \mathbf{Y}$, read 'X is distributed as Y', means that the distribution functions of **X** and **Y** are identical. All other associated functions, sets and probability statements of **X** and **Y** are therefore also identical.

'Is distributed as' is an equivalence relation, so that

(1) $\mathbf{X} \sim \mathbf{X}$
(2) $\mathbf{X} \sim \mathbf{Y} \Rightarrow \mathbf{Y} \sim \mathbf{X}$
(3) $\mathbf{X} \sim \mathbf{Y}$ and $\mathbf{Y} \sim \mathbf{Z} \Rightarrow \mathbf{X} \sim \mathbf{Z}$

We shall occasionally use the symbol $\approx$ meaning 'is approximately distributed as'. The symbol $\Rightarrow$ means 'implies'.

3.4.5 *Inverse function of a variate*

*Theorem 3.4*b: If **X** and **Y** are variates and ϕ is a one to one function then $\mathbf{Y} \sim \phi(\mathbf{X})$ implies $\phi^{-1}(\mathbf{Y}) \sim \mathbf{X}$.

Proof: $\mathbf{Y} \sim \phi(\mathbf{X}) \Rightarrow \text{Prob}\,[\mathbf{Y} \leqslant x] = \text{Prob}\,[\phi(\mathbf{X}) \leqslant x]$
(by Section 3.4.4)

$= \text{Prob}\,[\mathbf{X} \leqslant \phi^{-1}(x)]$
(by Section 3.3.2 and equation (3.4a)) (3.4f)

$\text{Prob}\,[\mathbf{Y} \leqslant x] = \text{Prob}\,[\phi^{-1}(\mathbf{Y}) \leqslant \phi^{-1}(x)]$
(by Section 3.3.2 and equation (3.4a)) (3.4g)

Equations (3.4f) and (3.4g) together with Section 3.4.4 imply that Theorem 3.4b is true. QED

3.5 Parameters

Every variate has an associated distribution function. Some groups of variates have distribution functions which differ from one another only in the values of certain parameters. A generalised distribution function in which the parameters appear as symbols corresponds to a family of variates. Examples are the variate families of the normal, lognormal, beta, gamma and exponential distributions. The detailed choice of the parameters which appear in a distribution function is to some extent arbitrary. However, we regard three types of parameter as 'basic' in the sense that they always have a certain physical or geometrical meaning. These are the location, scale and shape parameters the descriptions of which are as follows:

Location parameter, a: The abscissa of a location point (usually the lower or mid point) of the range of the variate.

Scale parameter, b: A parameter which determines the scale of measurement of the fractile, x.

Shape parameter, c: A parameter which determines the shape (in a sense distinct from location and scale) of the distribution function (and other functions) within a family of shapes associated with a specified type of variate.

The symbols a, b, c will be used to denote location, scale and shape parameters in general, but other symbols may be used in cases where firm conventions are established. Thus for the normal distribution the mean, μ, is a location parameter (the locating point is the mid point of the range) and the standard deviation, σ, is a scale parameter. The normal distribution does not have a shape parameter. Some distributions (for example the beta) have two shape parameters which we denote by v and w.

3.5.1 Variate and function notation

A variate **X** with parameters a, b, c is denoted in full by $\mathbf{X}: a, b, c$. Some or all of the parameters may be omitted if the context permits.

The distribution function for a variate $\mathbf{X}: c$ is $F_{\mathbf{X}}(x : c)$. If the variate name is implied by the context we write $F(x : c)$. Similar usages apply to other functions. The inverse distribution function for a variate $\mathbf{X}: a, b, c$ at probability level α is denoted $G_{\mathbf{X}}(\alpha : a, b, c)$.

3.6 Transformation of location and scale

Let $\mathbf{X}: 0, 1$ denote a variate with location parameter $a = 0$ and scale parameter $b = 1$. A variate which differs from $\mathbf{X}: 0, 1$ only in regard to location and scale is denoted $\mathbf{X}: a, b$ and is defined by

$$\mathbf{X}: a, b \sim a + b\,\mathbf{X}: 0, 1 \qquad (3.6a)$$

The location and scale transformation function is the one to one function

$$\phi(x) = a + bx \qquad (3.6b)$$

and its inverse is

$$\phi^{-1}(x) = (x - a)/b \qquad (3.6c)$$

The following equations relating to variates which differ only in relation to location and scale parameters then hold

$\mathbf{X}: a, b \sim a + b\,(\mathbf{X}: 0, 1)$
(by definition) (3.6a)

$\mathbf{X}: 0, 1 \sim [(\mathbf{X}.\, a, b) - a]/b$
(by Theorem 3.4b and equation (3.6a)) (3.6d)

$\text{Prob}\,[\mathbf{X}: a, b \leqslant x] = \text{Prob}\,[\mathbf{X}: 0, 1 \leqslant (x - a)/b]$
(by equation (3.4a)) (3.6e)

$F_{\mathbf{X}}(x : a, b) = F_{\mathbf{X}}\,\{[(x - a)/b] : 0, 1\}$
(equivalent to equation (3.6e)) (3.6f)

$G_{\mathbf{X}}(\alpha : a, b) \sim a + b\,G_{\mathbf{X}}(\alpha : 0, 1)$
(by Theorem 3.4a) (3.6g)

These and other interrelationships between functions associated with variates which differ only in regard to location and scale parameters are summarised in *Table 3.1.* The functions themselves are defined in *Table 2.1.*

Table 3.1 RELATIONSHIPS BETWEEN FUNCTIONS FOR VARIATES WHICH DIFFER ONLY BY LOCATION AND SCALE PARAMETERS a, b

Variate relationship	$\mathbf{X}: a, b \sim a + b(\mathbf{X}: 0, 1)$
Probability statement	$\text{Prob}\,[\mathbf{X}: a, b \leqslant x] = \text{Prob}\,[\mathbf{X}: 0, 1 \leqslant (x-a)/b]$
Function relationships:	
Distribution function	$F(x: a, b) = F([(x-a)/b]: 0, 1)$
Probability density function	$f(x: a, b) = (1/b)f([(x-a)/b]: 0, 1)$
Inverse distribution function	$G(\alpha: a, b) = a + bG(\alpha: 0, 1)$
Survival function	$S(x: a, b) = S([(x-a)/b]: 0, 1)$
Inverse survival function	$Z(\alpha: a, b) = a + Z(\alpha: 0, 1)$
Hazard function	$h(x: a, b) = (1/b)h([(x-a)/b]: 0, 1)$
Cumulative hazard function	$H(x: a, b) = H([(x-a)/b]: 0, 1)$
Moment generating function	$M(t: a, b) = \exp(at)M(bt: 0, 1)$
Laplace transform	$f^*(s: a, b) = \exp(-as)f^*(bs: 0, 1)$
Characteristic function	$C(t: a, b) = \exp(iat)C(bt: 0, 1)$
Cumulant function	$K(t: a, b) = iat + K(bt: 0, 1)$

3.7 Transformation from the rectangular variate

The following transformation is often useful for obtaining random numbers of a variate **X** from random numbers of the unit rectangular variate **R**. The latter has distribution function $F_{\mathbf{R}}(x) = x, 0 \leqslant x \leqslant 1$, and inverse distribution function $G_{\mathbf{R}}(\alpha) = \alpha, 0 \leqslant \alpha \leqslant 1$. The inverse distribution function of a general variate **X** is denoted $G_{\mathbf{X}}(\alpha), \alpha \in \mathcal{R}^{\alpha}{}_{\mathbf{X}}$. $G_{\mathbf{X}}(\alpha)$ is a one to one function.

Theorem 3.7a: $\mathbf{X} = G_{\mathbf{X}}(\mathbf{R})$

Proof: $\text{Prob}\,[\mathbf{R} \leqslant \alpha] = \alpha, 0 \leqslant \alpha \leqslant 1$

(property of **R**) (3.7a)

$= \text{Prob}\,[G_{\mathbf{X}}(\mathbf{R}) \leqslant G_{\mathbf{X}}(\alpha)]$

(by equation (3.4a)) (3.7b)

$\therefore \quad G_{\mathbf{X}}(\mathbf{R}) = \mathbf{X}$

(by equations (3.7a), (3.7b) and (2.3f)) QED

Thus every variate is related to the unit rectangular variate via its inverse distribution function, although of course this function will not always have a simple algebraic form.

3.8 Many to one transformations

In Sections 3.3 through 3.7 we considered the relationships between variates which were linked by a one to one function. Now we consider many to one functions which are defined as follows. Let ϕ be a function mapping from the real line into the real line.

Definition 3.8a: ϕ is many to one if there are at least two numbers x_1, x_2 in the domain of ϕ such that $\phi(x_1) = \phi(x_2)$, $x_1 \neq x_2$.

The many to one function $y = x^2$ is illustrated in *Figure 3.2*.

In Section 3.2 we defined, for a general variate $\mathbf{X}$ with range $\mathcal{R}_{\mathbf{X}}$ and for a function ϕ, a variate $\phi(\mathbf{X})$ with range $\mathcal{R}_{\phi(\mathbf{X})}$. $\phi(\mathbf{X})$ has the property that if $x_{\mathbf{X}}$ is a random number of $\mathbf{X}$ then $\phi(x_{\mathbf{X}})$ is a random number of $\phi(\mathbf{X})$. Let r_2 be a subset of $\mathcal{R}_{\phi(\mathbf{X})}$ and r_1 be the subset of $\mathcal{R}_{\mathbf{X}}$ which ϕ maps into r_2. The definition of $\phi(\mathbf{X})$ implies that

$$\text{Prob}\,[\mathbf{X} \in r_1] = \text{Prob}\,[\phi(\mathbf{X}) \in r_2] \tag{3.8a}$$

Relationships between $\mathbf{X}$ and $\phi(\mathbf{X})$ and their associated functions can be established using equation (3.8a). If ϕ is many to one the relationships will depend on the detailed form of ϕ.

3.8.1 Example

As an example we consider the relationships between the variates $\mathbf{X}$ and $\mathbf{X}^2$ for the case where $\mathcal{R}_{\mathbf{X}}$ is the real line. We know that $\phi : x \to x^2$ is a many to one function. In fact it is a two to one function in that $+x$ and $-x$ both map into x^2. Hence the probability that an as yet unrealised random number of $\mathbf{X}^2$ will be greater than x^2 will be equal to the probability that an as yet unrealised random number of $\mathbf{X}$ will be either greater than $+x$ or less than $-x$.

$$\text{Prob}\,[\mathbf{X}^2 > x^2] = \text{Prob}\,[\mathbf{X} > +x] + \text{Prob}\,[\mathbf{X} < -x] \tag{3.8b}$$

3.8.2 Symmetrical distributions

Let us now consider a variate $\mathbf{X}$ whose probability density function is symmetrical about the origin. We shall derive a relationship between the distribution functions of the variates $\mathbf{X}$ and $\mathbf{X}^2$ under the condition that $\mathbf{X}$ is symmetrical. An application of this result appears in the relationship between the **F** (variance ratio) and Student's **T** variates.

*Theorem 3.8.2*a

Let $\mathbf{X}$ be a variate whose probability density function is symmetrical about the origin.

(1) The distribution functions $F_{\mathbf{X}}(x)$ and $F_{\mathbf{X}^2}(x^2)$ for the variates $\mathbf{X}$ and $\mathbf{X}^2$ at fractiles x and x^2 respectively are related by

$$F_{\mathbf{X}}(x) = \tfrac{1}{2}\,[1 + F_{\mathbf{X}^2}(x^2)]$$

(2) The inverse survival functions $Z_{\mathbf{X}}(\frac{1}{2}\alpha)$ and $Z_{\mathbf{X}^2}(\alpha)$ for the variates $\mathbf{X}$ and $\mathbf{X}^2$ at probability levels $\frac{1}{2}\alpha$ and α respectively are related by

$$[Z_{\mathbf{X}}(\tfrac{1}{2}\alpha)]^2 = Z_{\mathbf{X}^2}(\alpha)$$

Proof:

(1) For a variate $\mathbf{X}$ with symmetrical p.d.f. we have

$$\text{Prob}\,[\mathbf{X} > x] = \text{Prob}\,[\mathbf{X} < x] \tag{3.8c}$$

Equations (3.8b) and (3.8c) imply

$$\text{Prob}\,[\mathbf{X}^2 > x^2] = 2\ \text{Prob}\,[\mathbf{X} > x] \tag{3.8d}$$

Introducing the distribution function $F_{\mathbf{X}}(x)$ we have, from the definition (equation (2.2a))

$$1 - F_{\mathbf{X}}(x) = \text{Prob}\,[\mathbf{X} > x] \tag{3.8e}$$

Equations (3.8d) and (3.8e) imply

$$1 - F_{\mathbf{X}^2}(x^2) = 2[1 - F_{\mathbf{X}}(x)] \tag{3.8f}$$

Rearrangement of equation (3.8f) gives

$$F_{\mathbf{X}}(x) = \tfrac{1}{2}\,[1 + F_{\mathbf{X}^2}(x^2)] \tag{3.8g}$$

(2) Let $$F_{\mathbf{X}}(x) = \alpha.$$

Equation (3.8g) implies

$$\tfrac{1}{2}\,[1 + F_{\mathbf{X}^2}(x^2)] = \alpha \tag{3.8h}$$

$$F_{\mathbf{X}^2}(x^2) = 2\alpha - 1 \quad \text{(rearrangement of equation (3.8h))} \tag{3.8i}$$

Equations (3.8i), (2.3a) and (2.3b) imply

$$G_{\mathbf{X}}(\alpha) = x \quad \text{and} \quad G_{\mathbf{X}^2}(2\alpha - 1) = x^2 \tag{3.8j}$$

Equation (3.8j) implies

$$[G_{\mathbf{X}}(\alpha)]^2 = G_{\mathbf{X}^2}(2\alpha - 1) \tag{3.8k}$$

From the definition of the inverse survival function, Z, (*Table 2.1*, item 6) we have $G(\alpha) = Z(1 - \alpha)$. Hence from equation (3.8k)

$$[Z_{\mathbf{X}}(1 - \alpha)]^2 = Z_{\mathbf{X}^2}(2(1 - \alpha))$$

$$[Z_{\mathbf{X}}(\alpha)]^2 = Z_{\mathbf{X}^2}(2\alpha)$$

$$[Z_{\mathbf{X}}(\alpha/2)]^2 = Z_{\mathbf{X}^2}(\alpha) \qquad \text{QED}$$

3.9 Functions of several variates

If **X** and **Y** are independent variates with ranges $\mathcal{R}_{\mathbf{X}}$ and $\mathcal{R}_{\mathbf{Y}}$ and ψ is a functional mapping from the Cartesian product of $\mathcal{R}_{\mathbf{X}}$ and $\mathcal{R}_{\mathbf{Y}}$ into the real line then $\psi(\mathbf{X}, \mathbf{Y})$ is a variate such that if $x_{\mathbf{X}}$ and $x_{\mathbf{Y}}$ are random numbers of **X** and **Y** respectively then $\psi(x_{\mathbf{X}}, x_{\mathbf{Y}})$ is a random number of $\psi(\mathbf{X}, \mathbf{Y})$.

The relationships between the associated functions of **X** and **Y** on the one hand and of $\psi(\mathbf{X}, \mathbf{Y})$ on the other are not generally straightforward and must be derived by analysis of the variates in question. One important general result is where the function is a summation, say $\mathbf{Z} = \mathbf{X} + \mathbf{Y}$. In this case practical results may often be obtained by using a property of the characteristic function $C_{\mathbf{X}}(t)$ of a variate **X**, namely, $C_{\mathbf{X}+\mathbf{Y}}(t) = C_{\mathbf{X}}(t) \,.\, C_{\mathbf{Y}}(t)$, i.e., the characteristic function of the sum of two variates is the product of the characteristic functions of the individual variates.

We are often interested in the sum (or other functions) of two or more variates which are independently and identically distributed. Thus consider the case $\mathbf{Z} \sim \mathbf{X} + \mathbf{Y}$ where $\mathbf{X} \sim \mathbf{Y}$. In this case we write

$$\mathbf{Z} \sim \mathbf{X}_1 + \mathbf{X}_2$$

or if there are n variates to be summed

$$Z \sim \sum_{i=1}^{n} X_i$$

Note that $X_1 + X_2$ is not the same as $2X_1$, even though $X_1 \sim X_2$. $X_1 + X_2$ is a variate for which a random number can be obtained by choosing a random number of **X** and then another independent random number of **X** and then adding the two. The latter is a variate for which a random number can be obtained by choosing a single random number of **X** and multiplying it by two.

4 BERNOULLI

A Bernoulli trial is a probabilistic experiment which can have one of two outcomes, success ($x = 1$) or failure ($x = 0$) and in which the probability of success is p. We refer to p as the Bernoulli probability parameter.

Variate $\mathbf{B} : 1, p$. The general binomial variate is $\mathbf{B} : n, p$.
Range $x \in \{0, 1\}$
Parameter p, the Bernoulli probability parameter, $0 < p < 1$
Distribution function $F(0) = 1 - p; F(1) = 1$
Probability function $f(0) = 1 - p; f(1) = p$

4.1 Random number generation

$\mathbf{R}$ is a unit rectangular variate and $\mathbf{B}: 1, p$ is a Bernoulli variate with probability parameter p.
$\mathbf{R} \leqslant p$ implies $\mathbf{B} : 1, p = 1$; $\mathbf{R} > p$ implies $\mathbf{B} : 1, p = 0$.

4.2 Curtailed Bernoulli trial sequences

The binomial, geometric, Pascal and negative binomial variates are based on sequences of Bernoulli trials which are curtailed in various ways, e.g. after n trials or x successes. We shall use the following terminology

p = Bernoulli probability parameter (probability of success at a single trial)
n = number of trials
x = number of successes
y = number of failures
Binomial variate, $\mathbf{B} : n, p$ = number of successes in n trials
Geometric variate, $\mathbf{G} : p$ = number of trials up to and including the first success
Pascal variate, $\mathbf{C} : x, p$ = number of trials up to and including the xth success
Negative binomial variate, $\mathbf{Y} : x, p$ = number of failures before the xth success

These variates are interrelated in the following ways

Binomial and geometric:

$$\mathbf{B}: n, p = x$$

where $x + 1$ is the smallest integer such that

$$\sum_{i=1}^{x+1} \mathbf{G}_i : p > n$$

Binomial and Pascal:

$$\text{Prob}\,[\mathbf{B}: n, p < x] = \text{Prob}\,[\mathbf{C}: x, p > n]$$

Binomial and negative binomial:

$$\text{Prob}\,[\mathbf{B}: n, p < x] = \text{Prob}\,[\mathbf{Y}: x, p > n - x]$$

Geometric and Pascal:

$$\sum_{i=1}^{x} \mathbf{G}_i : p \sim \mathbf{C}: x, p$$

$$\mathbf{G}: p \sim \mathbf{C}: 1, p$$

Geometric and negative binomial:

$$\sum_{i=1}^{x} \mathbf{G}_i : p \sim x + \mathbf{Y}: x, p$$

Pascal and negative binomial:

$$\mathbf{C}: x, p \sim x + \mathbf{Y}: x, p$$

5 BETA

Variate $\boldsymbol{\beta}: v, w$
Range $0 \leqslant x \leqslant 1$
Shape parameters $v > 0, w > 0$

Probability density function $x^{v-1}(1-x)^{w-1}/B(v, w)$

where $B(v, w)$ is the beta function with parameters v, w, given by

$$B(v, w) = \int_0^1 u^{v-1}(1-u)^{w-1}\, du$$

r^{th} moment about the origin $\prod_{i=0}^{r-1} (v+i)/(v+w+i)$

Mean $v/(v+w)$

Variance $vw/(v+w)^2(v+w+1)$

Mode $(v-1)/(v+w+2),\ v > 1, w > 1$

Coefficient of skewness $\dfrac{2(w-v)(v+w+1)^{1/2}}{(v+w+2)(vw)^{1/2}}$

Coefficient of kurtosis

$$\frac{3(v+w)(v+w+1)(v+1)(2w-v)}{vw(v+w+2)(v+w+3)} + \frac{v(v-w)}{v+w}$$

Coefficient of variation $\left[\dfrac{w}{v(v+w+1)}\right]^{1/2}$

Probability density function if v and w are integers $\dfrac{(v+w-1)!x^{v-1}(1-x)^{w-1}}{(v-1)!(w-1)!}$

Probability density function if range is $a \leqslant x \leqslant a+b$. Here a is a location parameter and b a scale parameter

$$\frac{1}{bB(v, w)}\left[\frac{x-a}{b}\right]^{v-1}\left[\frac{b-(x-a)}{b}\right]^{w-1}$$

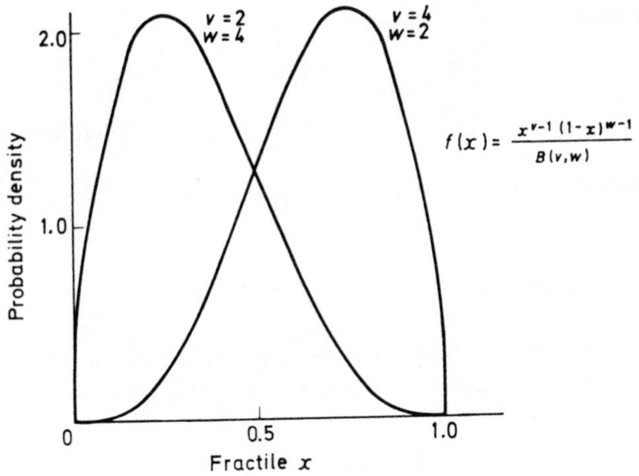

Figure 5.1 Probability density function for the beta variate **β**: *v, w*

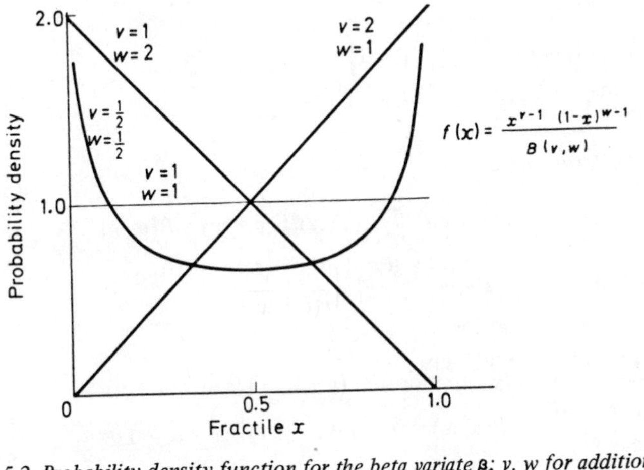

Figure 5.2 Probability density function for the beta variate **β**: *v, w for additional values of the parameters*

5.1 Parameter estimation

Parameter	*Estimator*	*Method*
v	$\bar{x}\,\{[\bar{x}(1-\bar{x})/s^2]-1\}$	Matching moments
w	$(1-\bar{x})\,\{[\bar{x}(1-\bar{x})/s^2]-1\}$	Matching moments

$\bar{x}$ = sample mean; s^2 = sample variance (unadjusted)

5.2 Notes on beta and gamma functions

The beta function with parameters v, w is denoted $B(v, w)$. The gamma function with parameter c is denoted $\Gamma(c)$; $v, w, c > 0$.

Definitions

Beta function: $$B(v, w) = \int_0^1 u^{v-1}(1-u)^{w-1}\,du$$

Gamma function: $$\Gamma(c) = \int_0^\infty \exp(-u)\,u^{c-1}\,du$$

Interrelationships

$$B(v, w) = \Gamma(v)\,\Gamma(w)/\Gamma(v+w) = B(w, v)$$

$$\Gamma(c) = (c-1)\,\Gamma(c-1)$$

Special Values
If v, w and c are integers

$$B(v, w) = (v-1)!\,(w-1)!/(v+w-1)!$$

$$\Gamma(c) = (c-1)!$$

$$B(1, 1) = 1,\ B(\tfrac{1}{2}, \tfrac{1}{2}) = \pi$$

$$\Gamma(0) = 1,\ \Gamma(1) = 1,\ \Gamma(\tfrac{1}{2}) = \pi^{1/2}$$

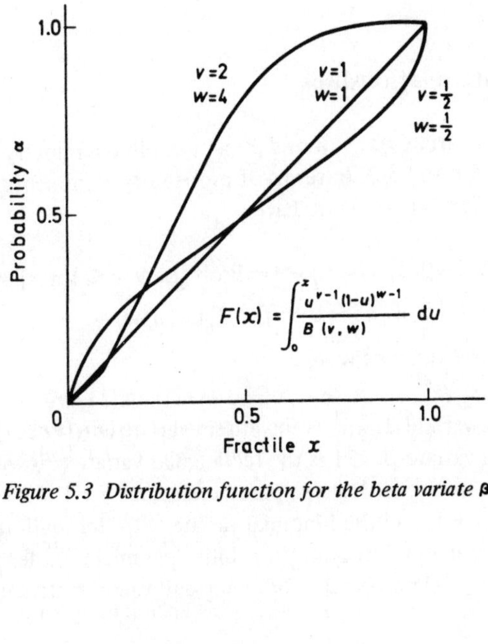

Figure 5.3 Distribution function for the beta variate β: v, w

Alternative Expressions

$$B(v, w) = 2\int_0^{\pi/2} \sin^{2v-1}(\theta)\cos^{2w-1}(\theta)\,d\theta$$

$$= \int_0^{\infty} \frac{y^{w-1}\,dy}{(1+y)^{v+w}}$$

5.3 Variate relationships

1. The beta variates $\boldsymbol{\beta}: v, w$ and $\boldsymbol{\beta}: w, v$ exhibit symmetry, *see Figures 5.1* and *5.2.* In terms of probability statements and the distribution functions we have

$$\text{Prob}\,[\boldsymbol{\beta}: v, w \leqslant x] = 1 - \text{Prob}\,[\boldsymbol{\beta}: w, v \leqslant 1 - x] =$$
$$\text{Prob}\,[\boldsymbol{\beta}: w, v > (1 - x)] = F_{\boldsymbol{\beta}}(x : v, w) =$$
$$1 - F_{\boldsymbol{\beta}}(1 - x : w, v)$$

2. The beta variate $\boldsymbol{\beta}: \frac{1}{2}, \frac{1}{2}$ is the arc sin variate (*Figure 5.1*).
3. The beta variate $\boldsymbol{\beta}: 1, 1$ is the rectangular variate (*Figure 5.2*).
4. The beta variate with shape parameters $i, n - i + 1$, denoted $\boldsymbol{\beta}: i, n - i + 1$, and the binomial variate with Bernoulli trial parameter n and Bernoulli probability parameter x, denoted $\mathbf{B}: n, x$, are related by the following equivalent statements

$$\text{Prob}\,[(\boldsymbol{\beta}: i, n - i + 1) \leqslant x] = \text{Prob}\,[(\mathbf{B}: n, x) \geqslant i]$$
$$F_{\boldsymbol{\beta}}(x : i, n - i + 1) = 1 - F_{\mathbf{B}}(i - 1 : n, x)$$

n and i are positive integers, $0 \leqslant x \leqslant 1$.

Equivalently, putting $v = i$ and $w = n - i + 1$

$$F_{\boldsymbol{\beta}}(x : v, w) = 1 - F_{\mathbf{B}}(v - 1 : v + w - 1, x)$$
$$= F_{\mathbf{B}}(w - 1 : v + w - 1, 1 - x)$$

5. The beta variate with shape parameters $w/2$, $v/2$, denoted $\boldsymbol{\beta}: w/2, v/2$, and the **F** variate with degrees of freedom v, w, denoted $\mathbf{F}: v, w$, are related by

$$\text{Prob}\,[(\boldsymbol{\beta}: w/2, v/2) \leqslant [w/(w+vx)]]$$
$$= \text{Prob}\,[\mathbf{F}: v, w > x]$$

Hence the inverse distribution function $G_{\boldsymbol{\beta}}(\alpha : w/2, v/2)$ of the beta variate $\boldsymbol{\beta}: w/2, v/2$ and the inverse survival function $Z_{\mathbf{F}}(\alpha : v, w)$ of the **F** variate $\mathbf{F}: v, w$ are related by

$$(w/v)\,\{[1/G_{\boldsymbol{\beta}}(\alpha : w/2, v/2)] - 1\} = Z_{\mathbf{F}}(\alpha : v, w)$$
$$= G_{\mathbf{F}}(1-\alpha : v, w)$$

where α denotes probability.

6. The gamma variate with shape parameter v, denoted $\boldsymbol{\gamma}: v$, and the gamma variate with shape parameter w, denoted $\boldsymbol{\gamma}: w$, are related to the beta variate $\boldsymbol{\beta}: v, w$ by

$$\boldsymbol{\beta}: v, w \sim (\boldsymbol{\gamma}: v)/(\boldsymbol{\gamma}: v + \boldsymbol{\gamma}: w)$$

where $\boldsymbol{\gamma}: v$, $\boldsymbol{\gamma}: w$, $\boldsymbol{\beta}: v, w$ all have unit scale parameter.

5.4 Random number generation

If v and w are integers then random numbers of the beta variate $\boldsymbol{\beta}: v, w$ can be computed from random numbers of the unit rectangular variate **R** using the relationship between $\boldsymbol{\beta}: v, w$ and the gamma variates $\boldsymbol{\gamma}: v$ and $\boldsymbol{\gamma}: w$ as follows

$$\boldsymbol{\gamma}: v \sim -\log \prod_{i=1}^{v} \mathbf{R}_i$$

$$\boldsymbol{\gamma}: w \sim -\log \prod_{j=1}^{w} \mathbf{R}_j$$

$$\boldsymbol{\beta}: v, w \sim \boldsymbol{\gamma}: v/(\boldsymbol{\gamma}: v + \boldsymbol{\gamma}: w)$$

6 BINOMIAL

Variate **B** : n, p
Fractile x, number of successes
Range $0 \leqslant x \leqslant n$, x an integer
The binomial variate **B** : n, p is the number of successes in n independent Bernoulli trials where the probability of success at each trial is p and the probability of failure is $q = 1 - p$.
Parameters n, the Bernoulli trial parameter, n a positive integer
p, the Bernoulli probability parameter, $0 < p < 1$

Distribution function	$\sum_{i=0}^{x} \binom{n}{i} p^i q^{n-i}$
Probability function	$\binom{n}{x} p^x q^{n-x}$
Moment generating function	$[p \exp(t) + q]^n$
Probability generating function	$(pt + q)^n$
Characteristic function	$[p \exp(\mathrm{i}t) + q]^n$
Moments about the origin:	
Mean	np
Second	$np(np + q)$
Third	$np[(n-1)(n-2)p^2 + 3p(n-1) + 1]$
Moments about the mean:	
Variance	npq
Third	$npq(q - p)$
Fourth	$npq[1 + 3pq(n-2)]$
Standard deviation	$(npq)^{1/2}$
Mode	$p(n+1) - 1 \leqslant x \leqslant p(n+1)$
Coefficient of skewness	$(q-p)/(npq)^{1/2}$
Coefficient of kurtosis	$3 - \frac{6}{n} + \frac{1}{npq}$

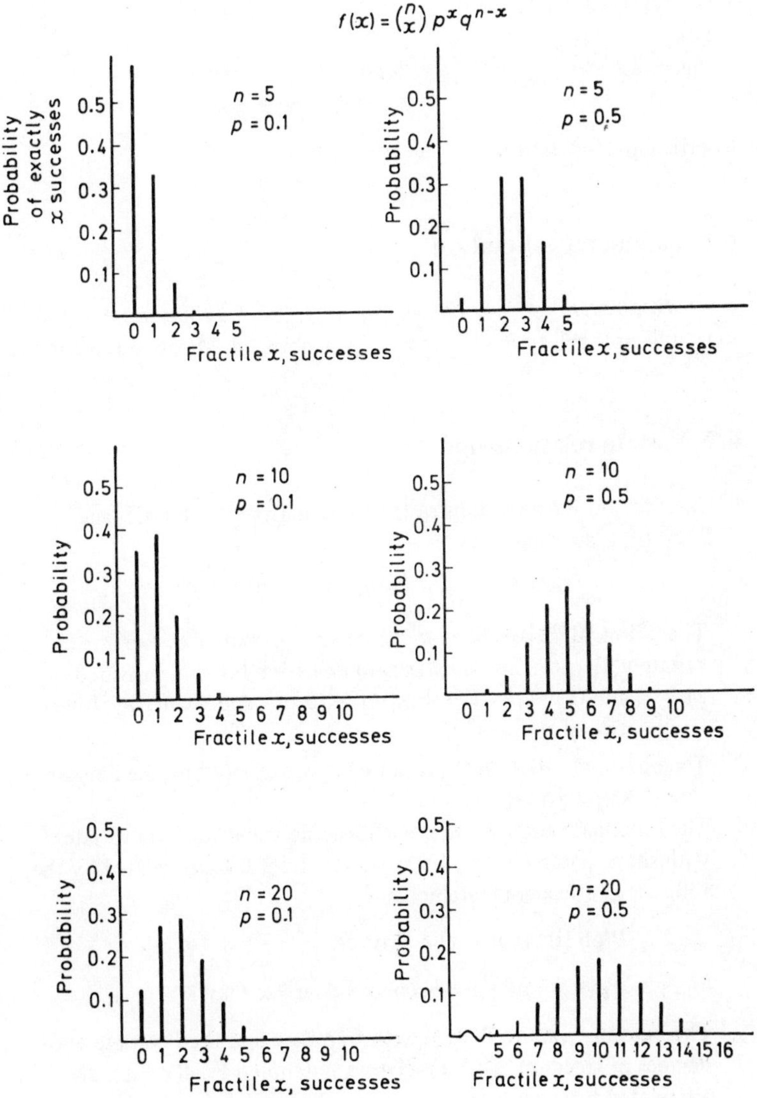

Figure 6.1 Probability function for the binomial variate **B**: *n, p*

Factorial moments about the mean:	
Second	npq
Third	$-2npq(1+p)$
Coefficient of variation	$(q/np)^{1/2}$

6.1 Parameter estimation

Parameter	*Estimator*	*Properties*
Bernoulli probability, p	x/n	Unbiassed, Maximum likelihood

6.2 Variate relationships

1. The distribution functions of the binomial variates $\mathbf{B}: n, p$ and $\mathbf{B}: n, 1-p$ are related by

$$F_{\mathbf{B}}(x: n, p) = 1 - F_{\mathbf{B}}(n-x-1: n, 1-p)$$

2. The binomial variate $\mathbf{B}: n, p$ can be approximated by the normal variate with mean np and standard deviation $(np)^{1/2}$, provided $np > 5$ and $0.1 \leqslant p \leqslant 0.9$. For $np > 25$ this approximation holds irrespective of p.
3. The binomial variate $\mathbf{B}: n, p$ can be approximated by the Poisson variate with mean np provided $p < 0.1$.
4. The binomial variate $\mathbf{B}: n, p$ with fractile x and the beta variate with shape parameters $x, n-x+1$ and fractile p are related by the following equivalent statements

$$\text{Prob}\,[\mathbf{B}: n, p \geqslant x] = \text{Prob}\,[\boldsymbol{\beta}: x, n-x+1 \leqslant p]$$

$$F_{\mathbf{B}}(x: n, p) = 1 - F_{\boldsymbol{\beta}}(p: x+1, n-x+2)$$

5. The binomial variate $\mathbf{B}: n, p$ with fractile x and the $\mathbf{F}$ variate with degrees of freedom $2(x+1), 2(n-x)$, denoted $\mathbf{F}: 2(x+1), 2(n-x)$, are related by

$$\text{Prob}\,[\mathbf{B}: n, p \leqslant x] = 1 - \text{Prob}\,[\mathbf{F}: 2(x+1), 2(n-x) < p(n-x)/(1+x)(1-p)]$$

$$F(x) = \sum_{i=0}^{x} \binom{n}{i} p^i q^{n-i}$$

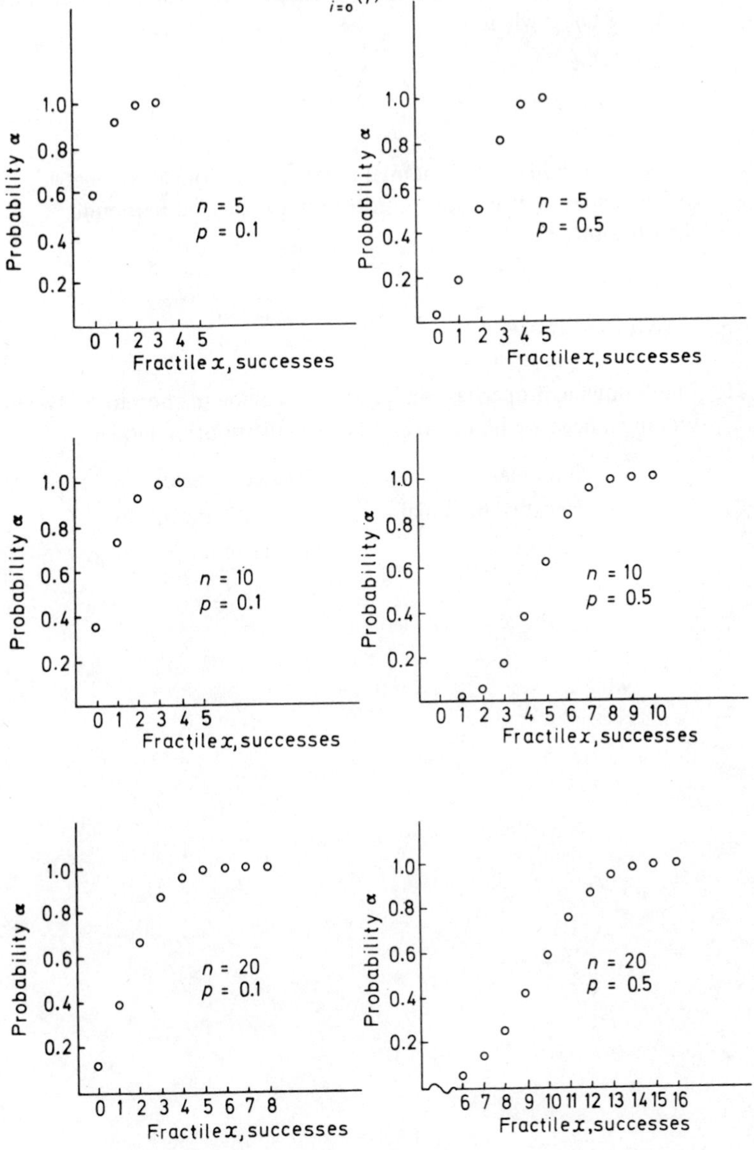

Figure 6.2 Distribution function for the binomial variate **B**: *n, p*

6. The sum of k binomial variates $\mathbf{B}: n_i, p; i = 1 \ldots k$ is the binomial variate $\mathbf{B}: n', p$ where

$$n' = \sum_{i=1}^{k} n_i$$

7. The relationship of the binomial to negative binomial, geometric and Pascal variates is detailed in the section headed Bernoulli distribution.

6.3 Note

1. The following properties can be used as a guide in choosing between binomial, negative binomial and Poisson distribution models

Binomial	Variance < mean
Negative binomial	Variance > mean
Poisson	Variance = mean

6.4 Random number generation

1. *Rejection technique:* Select n unit rectangular random numbers. The number of these which are less than p is a random number of the binomial variate $\mathbf{B}: n, p$.
2. *Geometric distribution method:* If p is small a method which may be faster than the rejection technique is to add together geometric random numbers until their sum exceeds n. The number of geometric random numbers minus one is a binomial random number.
   $\mathbf{B}: n, p = k - 1$ where k is the smallest integer such that

$$\sum_{i=1}^{k} \mathbf{G}_i : p > n$$

$\mathbf{G}_i : p = \log(\mathbf{R}_i)/\log(1 - p)$ rounded up to the next larger integer

where $\mathbf{R}$ is a unit rectangular variate.

7 CAUCHY

Range $-\infty \leqslant x \leqslant +\infty$	
Location parameter a, the median	
Scale parameter b	
Probability density function	$1/\pi b\{[(x-a)/b]^2+1\}$
Moments about the median	Do not exist
Mode	a
Median	a

7.1 Variate relationships

1. The ratio of two independent unit normal variates $\mathbf{N}_1$, $\mathbf{N}_2$ is a Cauchy variate with zero median and unit scale parameter, denoted $\mathbf{X}: 0, 1$.

$$\mathbf{N}_1/\mathbf{N}_2 \sim \mathbf{X}: 0, 1$$

2. The sum of n Cauchy variates $\mathbf{X}_i : a_i, b_i$ with location parameters $a_i, i = 1, \ldots, n$ and scale parameters $b_i, i = 1, \ldots, n$ is a Cauchy variate $\mathbf{X}: a, b$ with parameters the sum of those of the individual variates

$$\sum_{i=1}^{n} \mathbf{X}_i : a_i, b_i \sim \mathbf{X}: a, b$$

$$a = \sum_{i=1}^{n} a_i; \; b = \sum_{i=1}^{n} b_i$$

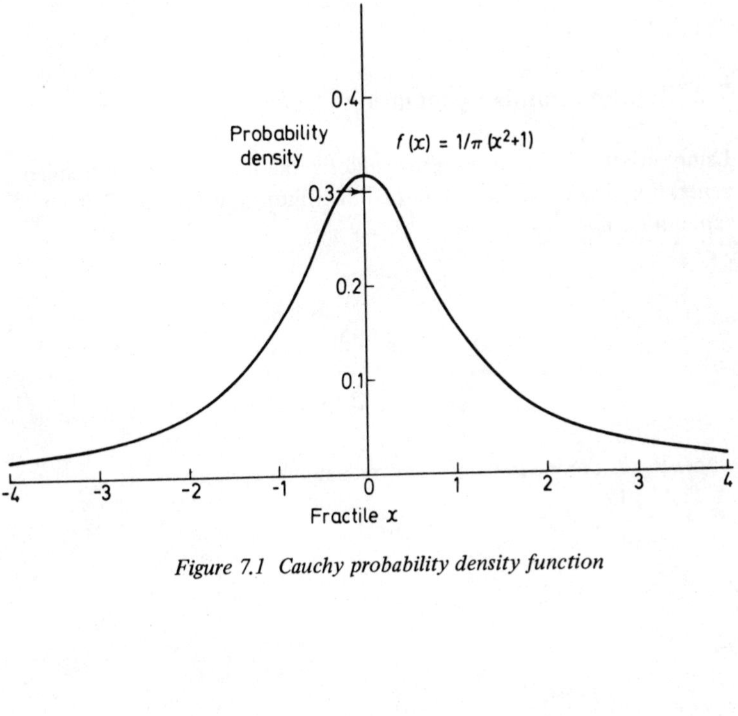

Figure 7.1 Cauchy probability density function

3. The reciprocal of a Cauchy variate $\mathbf{X}: a, b$ is a Cauchy variate $\mathbf{X}: a', b'$ where a', b' are as shown below

$$1/\mathbf{X}: a, b \sim \mathbf{X}: a', b'$$
$$a' = a/(a^2 + b^2); \; b' = b/(a^2 + b^2)$$

7.2 Random number generation

Using variate relationship 1 above and the method of random number generation described for the normal distribution, we obtain Cauchy random numbers as follows

$$\mathbf{X}: 0, 1 \approx \frac{\sum_{i=1}^{12} \mathbf{R}_i - 6}{\sum_{j=1}^{12} \mathbf{R}_j - 6},$$

where $\mathbf{R}_i$, $\mathbf{R}_j$ are independent unit rectangular variates, $i = 1, \ldots, 12$; $j = 1, \ldots, 12$.

7.3 General form

General form of the probability density function with shape parameter m, normalising constant k, and zero location and unit scale parameter	$f(x) = k/(1 + x^2)^m, \; m \geqslant 1$ where $k = \Gamma(m)/\Gamma(\frac{1}{2})\Gamma(m - \frac{1}{2})$
$2r^{\text{th}}$ moment about the origin	$\Gamma(r + \frac{1}{2})\Gamma(m - r - \frac{1}{2})/\Gamma(\frac{1}{2})\Gamma(m - \frac{1}{2}), \; 2m > 2r + 1$

7.4 Notes

1. Symmetrical about $x = 0$.
2. Odd moments about the origin are zero.
3. Moments about the origin of order $2r < 2m - 1$ do not exist.
4. Mode at $x = 0$.

8 CHI-SQUARED

Variate $\chi^2 : v$
Range $0 \leqslant x \leqslant +\infty$
Shape parameter v, degrees of freedom

Probability density function	$\dfrac{x^{(v-2)/2} \exp(-x/2)}{2^{v/2}\,\Gamma(v/2)}$ where $\Gamma(v/2)$ is the gamma function with parameter $v/2$
Moment generating function	$(1-2t)^{-v/2},\ t > \frac{1}{2}$
Laplace transform of the p.d.f.	$(1+2s)^{-v/2}$
Characteristic function	$(1-2it)^{-v/2}$
Cumulant function	$(-v/2)\log(1-2it)$
r^{th} cumulant	$2^{r-1}v[(r-1)!]$
r^{th} moment about the origin	$2^r \prod_{i=0}^{r-1} [i+(v/2)]$
Mean	v
Variance	$2v$
Standard deviation	$(2v)^{1/2}$
Mode	$v-2,\ v \geqslant 2$
Coefficient of skewness	$2^{3/2}v^{-1/2}$
Coefficient of kurtosis	$3+12/v$
Coefficient of variation	$(2/v)^{1/2}$

8.1 Variate relationships

1. The chi-squared variate with v degrees of freedom is equal to the gamma variate with scale parameter 2 and shape parameter $v/2$, or

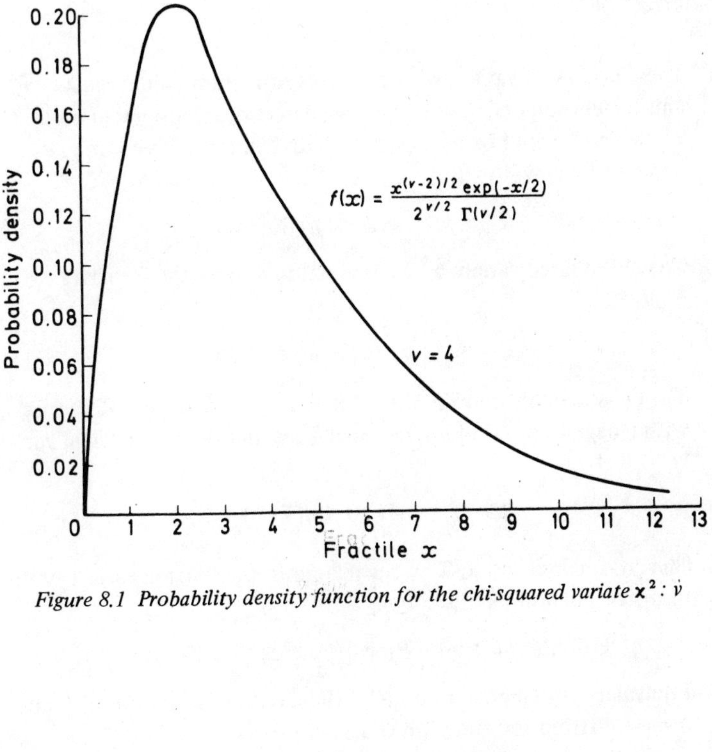

Figure 8.1 Probability density function for the chi-squared variate $\chi^2 : \nu$

equivalently is twice the gamma variate with scale parameter 1 and shape parameter $v/2$.

$$\boldsymbol{\chi}^2 : v \sim \boldsymbol{\gamma} : 2,\ v/2 \sim 2(\boldsymbol{\gamma} : 1, v/2)$$

Properties of the gamma variate $\boldsymbol{\gamma} : 2, v/2$ apply to the chi-squared variate $\boldsymbol{\chi}^2 : v$.

2. The chi-squared variate with v degrees of freedom, denoted $\boldsymbol{\chi}^2 : v$, and the chi-squared variate with w degrees of freedom, denoted $\boldsymbol{\chi}^2 : w$, are related to the **F** variate with degrees of freedom v, w, denoted $\mathbf{F} : v, w$, by

$$\mathbf{F} : v, w \sim w(\boldsymbol{\chi}^2 : v)/v(\boldsymbol{\chi}^2 : w)$$

3. The chi-squared variate $\boldsymbol{\chi}^2 : v$ is equal to v times the **F** variate $\mathbf{F} : v, \infty$.

$$\boldsymbol{\chi}^2 : v \sim v(\mathbf{F} : v, \infty)$$

4. The chi-squared variate $\boldsymbol{\chi}^2 : v$ is related to the Student's **T** variate with v degrees of freedom, denoted $\mathbf{T} : v$, and the unit normal variate $\mathbf{N} : 0, 1$, by

$$\boldsymbol{\chi}^2 : v \sim v(\mathbf{N} : 0, 1)/(\mathbf{T} : v)$$

5. The chi-squared variate $\boldsymbol{\chi}^2 : v$ is related to the Poisson variate with mean $x/2$, denoted $\mathbf{P} : x/2$, by

$$\text{Prob}\,[\boldsymbol{\chi}^2 : v > x] = \text{Prob}\,[(\mathbf{P} : x/2) \leqslant (v/2) - 1]$$

Equivalent statements in terms of the distribution functions F and inverse distribution function G are

$$1 - F_{\boldsymbol{\chi}^2}(x : v) = F_{\mathbf{P}}([(v/2) - 1] : x/2)$$

$$G_{\boldsymbol{\chi}^2}((1 - \alpha) : v) = x \Leftrightarrow G_{\mathbf{P}}(\alpha : x/2) = (v/2) - 1$$

$0 \leqslant x \leqslant +\infty$; $v/2$ a positive integer; $0 \leqslant \alpha \leqslant 1$; α denotes probability.

6. The chi-squared variate $\boldsymbol{\chi}^2 : v$ is equal to the sum of the squares of v unit normal variates.

$$\boldsymbol{\chi}^2 : v \sim \sum_{i=1}^{v} (\mathbf{N}_i : 0, 1)^2 \sim \sum_{i=1}^{v} \{[(\mathbf{N}_i : \mu, \sigma) - \mu]/\sigma\}^2$$

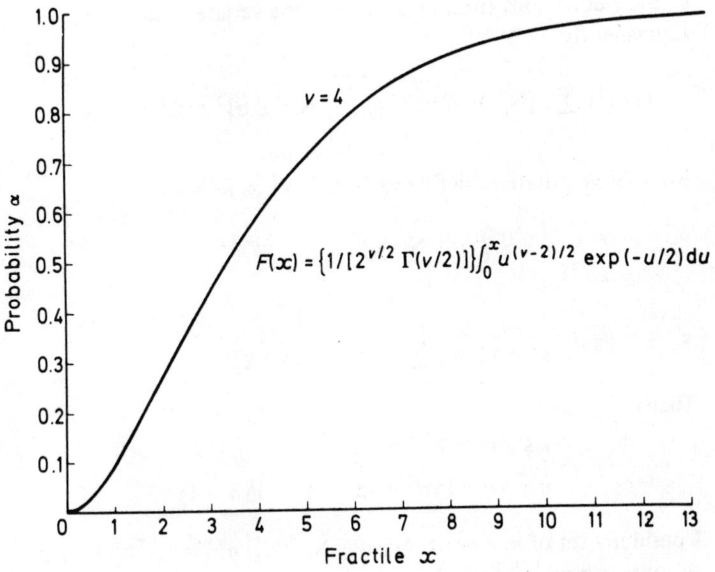

Figure 8.2 Distribution function for the chi-squared variate $\chi^2 : \nu$

7. The chi-squared variate $\chi^2 : \upsilon$ with $\upsilon > 30$ is approximately equal to half the square of a normal variate with mean $(2\upsilon - 1)^{1/2}$ and unit variance, $\mathbf{N} : (2\upsilon - 1)^{1/2}, 1$.

$$\chi^2 : \upsilon \approx \tfrac{1}{2} [\mathbf{N} : (2\upsilon - 1)^{1/2}, 1]^2 \sim \tfrac{1}{2} [(2\upsilon - 1)^{1/2} + \mathbf{N} : 0, 1]^2, \upsilon > 30$$

8. Given n normal variates $\mathbf{N}_i : \mu, \sigma; i = 1, \ldots, n$, the sum of the squares of their deviations from their mean is the variate $\sigma^2 \chi^2 : n - 1$. Equivalently

$$(1/\sigma^2) \sum_{i=1}^{n} [\mathbf{N}_i : \mu, \sigma - (1/n) \sum_{i=1}^{n} \mathbf{N}_i : \mu, \sigma]^2 \sim \chi^2 : n - 1$$

Alternative notation, define variates $\bar{\mathbf{x}}, \mathbf{s}^2$ as follows

$$\bar{\mathbf{x}} \sim (1/n) \sum_{i=1}^{n} \mathbf{N}_i : \mu, \sigma$$

$$\mathbf{s}^2 \sim (1/n) \sum_{i=1}^{n} [(\mathbf{N}_i : \mu, \sigma) - \bar{\mathbf{x}}]^2$$

Then

$$n\mathbf{s}^2/\sigma^2 \sim \chi^2 : n - 1$$

$$n\mathbf{s}^2/(n-1)\sigma^2 \sim (\chi^2 : n - 1)/(n - 1)$$

9. Consider a set of n_1 normal variates $\mathbf{N}_{1i} : \mu_1, \sigma; i = 1, \ldots, n_1$ and a set of n_2 normal variates $\mathbf{N}_{2j} : \mu_2, \sigma; j = 1, \ldots, n_2$ (note same S.D.) and define variates $\bar{\mathbf{x}}_1, \bar{\mathbf{x}}_2, \mathbf{s}_1^2, \mathbf{s}_2^2$ as follows

$$\bar{\mathbf{x}}_1 \sim (1/n_1) \sum_{i=1}^{n_1} \mathbf{N}_{1i} : \mu_1, \sigma$$

$$\bar{\mathbf{x}}_2 \sim (1/n_2) \sum_{j=1}^{n_2} \mathbf{N}_{2j} : \mu_2, \sigma$$

$$\mathbf{s}_1^2 \sim (1/n_1) \sum_{i=1}^{n_1} (\mathbf{N}_{1i} : \mu_1, \sigma - \bar{\mathbf{x}}_1)^2$$

$$\mathbf{s}_2^2 \sim (1/n_2) \sum_{j=1}^{n_2} (\mathbf{N}_{2j} : \mu_2, \sigma - \bar{\mathbf{x}}_2)^2$$

Then

$$(n_1 s_1^2 + n_2 s_2^2)/\sigma^2 \sim \boldsymbol{\chi}^2 : n_1 + n_2 - 2$$

8.2 Random number generation

$\boldsymbol{\chi}^2 : v$ denotes a chi-squared variate with v degrees of freedom
$\mathbf{R} : 0, 1$ denotes a unit rectangular variate
$\mathbf{N} : 0, 1$ denotes a unit normal variate
For v even

$$\boldsymbol{\chi}^2 : v \sim -\tfrac{1}{2} \log \left(\prod_{i=1}^{v/2} \mathbf{R}_i : 0, 1 \right)$$

For v odd

$$\boldsymbol{\chi}^2 : v \sim -\tfrac{1}{2} \log \left(\prod_{i=1}^{(v-1)/2} \mathbf{R}_i : 0, 1 \right) + (\mathbf{N} : 0, 1)^2$$

9 DISCRETE UNIFORM

Variate **D** : a, b
Range $a \leqslant x \leqslant a + b - 1$, x an integer
Location parameter a, the lower limit of the range
Scale parameter b
Alternative parameter h, the upper limit of the range, $h = a + b - 1$

Distribution function	$(x - a + 1)/b$
Probability function	$1/b$
Inverse distribution function (of probability α)	$a - 1 + \alpha b$
Survival function	$a + b - x - 1$
Inverse survival function (of probability α)	$a - 1 + (1 - \alpha)b$
Hazard function $h(x) = f(x + 1)/[1 - F(x)]$	$1/(a + b - x - 1)$
Probability generating function	$(t^a - t^{a+b})/(1 - t)$
Characteristic function	$\exp[i(a - 1)t] \sinh(itb/2)/b \sinh(it/2)$
Mean	$a + (b - 1)/2$
Variance	$(b^2 - 1)/12$
Coefficient of skewness	0
Information content	$\log_2 b$

9.1 Parameter estimation

Parameter	*Estimator*	*Method*
Location parameter, a	$\bar{x} - [(12s^2 + 1)^{1/2} - 1]/2$	Matching moments
Scale parameter, b	$(12s^2 + 1)^{1/2}$	Matching moments

$\bar{x}$ = sample mean; s^2 = sample variance (unadjusted)

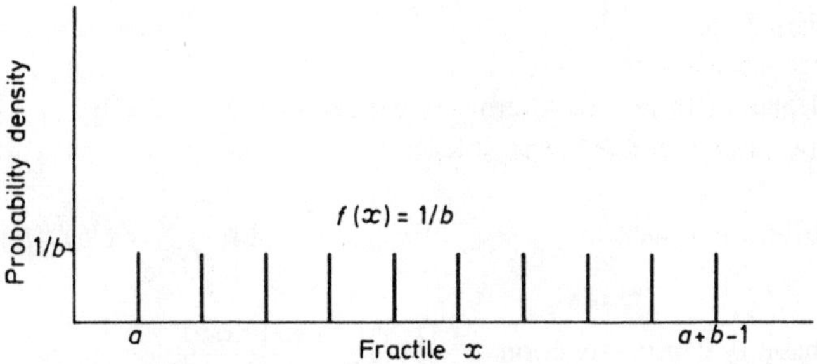

Figure 9.1 Probability density function for the discrete uniform variate **D**: *a, b*

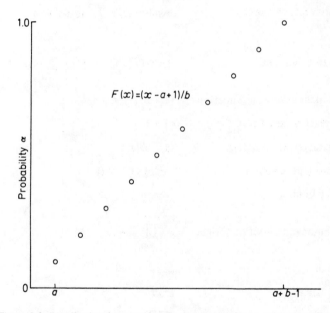

Figure 9.2 Distribution function for the discrete uniform variate **D**: *a, b*

The Erlang variate is a gamma variate with shape parameter c an integer. The diagrams, notes on parameter estimation and variate relationships for the gamma variate apply to the Erlang variate.

Variate $\gamma : b, c$
Range $0 \leqslant x \leqslant +\infty$
Scale parameter $b > 0$. Alternative parameter λ, $\lambda = 1/b$
Shape parameter $c > 0$, c an integer

Distribution Function	$1 - [\exp(-x/b)]\left[\sum_{i=0}^{c-1} (x/b)^i/(i!)\right]$
Probability density function	$\dfrac{(x/b)^{c-1}\exp(-x/b)}{b[(c-1)!]}$
Survival function	$\exp(-x/b)\left[\sum_{i=0}^{c-1} (x/b)^i/i!\right]$
Hazard function	$\{(x/b)^{c-1}/b[(c-1)!]\}\left[\sum_{i=0}^{c-1} (x/b)^i/(i!)\right]$
Moment generating function	$(1-bt)^{-c}, t > 1/b$
Laplace transform of the p.d.f.	$(1+bs)^{-c}$
Characteristic function	$(1-ibt)^{-c}$
Cumulant function	$-c\log(1-ibt)$
r^{th} cumulant	$(r-1)!\,cb^r$
r^{th} moment about the origin	$b^r \prod_{i=0}^{r-1} (c+i)$
Mean	bc
Variance	$b^2 c$
Standard deviation	$bc^{1/2}$
Mean deviation	$2\exp(-c)c^{c+1}/(c!)$
Mode	$b(c-1)$

Coefficient of skewness	$2c^{-1/2}$
Coefficient of kurtosis	$3 + 6/c$
Coefficient of variation	$c^{-1/2}$

10.1 Parameter estimation

See gamma distribution.

10.2 Variate relationships

1. If $c = 1$ the Erlang reduces to the exponential distribution.
2. The Erlang variate with scale parameter b and shape parameter c, denoted $\mathbf{\Upsilon}: b, c$, is equal to the sum of c exponential variates with mean b, denoted $\mathbf{E}_i : b; i = 1, \ldots, c$.

$$\mathbf{\Upsilon}: b, c \sim \sum_{i=1}^{c} \mathbf{E}_i : b, \quad c \text{ a positive integer}$$

3. For other properties see the gamma distribution.

10.3 Random number generation

$$\mathbf{\Upsilon}: b, c \sim -b \log\left[\prod_{i=1}^{c} \mathbf{R}_i\right]$$

where $\mathbf{R}_i$ is a rectangular variate with range 0, 1.

11 EXPONENTIAL

Variate **E** : b
Range $0 \leqslant x \leqslant +\infty$
Scale parameter b, the mean
Alternative parameter λ, the hazard function (hazard rate), $\lambda = 1/b$

Distribution function	$1 - \exp(-x/b)$
Probability density function	$(1/b) \exp(-x/b)$
Inverse distribution function (of probability α)	$b \log [1/(1 - \alpha)]$
Survival function	$\exp(-x/b)$
Inverse survival function (of probability α)	$b \log (1/\alpha)$
Hazard function (Hazard rate)	$1/b$
Cumulative hazard function	x/b
Moment generating function	$1/(1 - bt),\ t > (1/b)$
Laplace transform of the p.d.f.	$1/(1 + bs)$
Characteristic function	$1/(1 - \mathrm{i}bt)$
Cumulant function	$-\log (1 - \mathrm{i}bt)$
r^{th} cumulant	$(r - 1)!b^r$
r^{th} moment about the origin	$r!b^r$
Mean	b
Variance	b^2
Standard deviation	b
Mean deviation	$2b/\mathrm{e}$ where e is the base of natural logarithms
Mode	0
Median	$b \log 2$
Coefficient of skewness	2

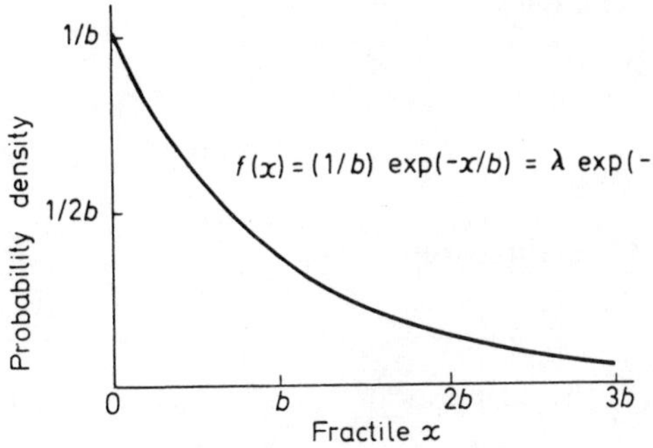

Figure 11.1 Probability density function for the exponential variate E: *b*

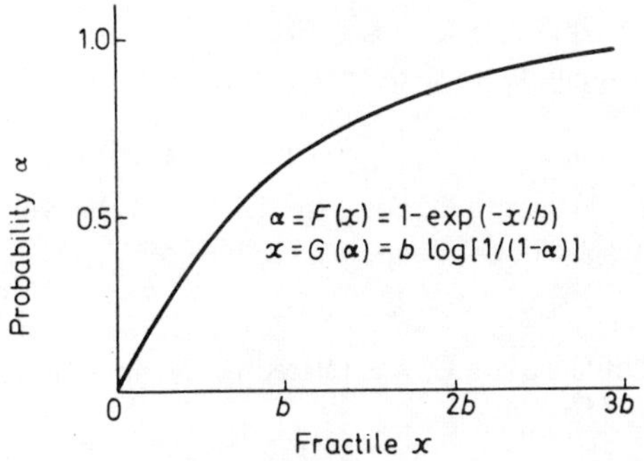

Figure 11.2 Distribution function for the exponential variate E: *b*

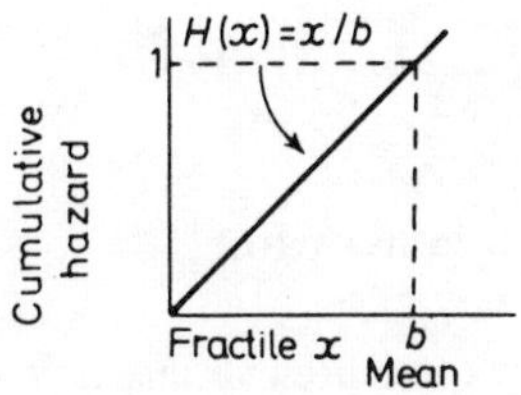

Figure 11.3 Cumulative hazard function for the exponential variate E: *b. For method of use see Wayne Nelson, 'Hazard Plotting for Incomplete Failure Data', Journal of Quality Technology,* 1, *No. 1, Jan. 1969*

Coefficient of kurtosis	9
Coefficient of variation	1
Information content	$\log_2 eb$ where e is the base of natural logarithms

11.1 Parameter estimation

Parameter	*Estimator*	*Properties*
Mean, b	$\bar{x}$, the sample mean	Unbiassed, Maximum likelihood

11.2 Variate relationships

1. The exponential variate $\mathbf{E}: b$ is a special case of the gamma variate $\boldsymbol{\gamma}: b, c$ corresponding to shape parameter $c = 1$.

$$\mathbf{E}: b \sim \boldsymbol{\gamma}: b, 1$$

2. The exponential variate $\mathbf{E}: b$ is a special case of the Weibull variate $\mathbf{W}: b, c$ corresponding to shape parameter $c = 1$.

$$\mathbf{E}: b \sim \mathbf{W}: b, 1$$

3. The exponential variate $\mathbf{E}: b$ is related to the uniform variate $\mathbf{R}$ by

$$\mathbf{E}: b \sim -b \log \mathbf{R}$$

4. The sum of n exponential variates $\mathbf{E}_i : b; i = 1, \ldots, n$, is the Erlang (gamma) variate $\boldsymbol{\gamma}: b, n$.

$$\sum_{i=1}^{n} \mathbf{E}_i : b \sim \boldsymbol{\gamma}: b, n$$

11.3 Random number generation

Random numbers of the exponential variate $\mathbf{E}: b$ can be computed from random numbers of the unit rectangular variate $\mathbf{R}$ using the relationship

$$\mathbf{E}: b \sim -b \log \mathbf{R}$$

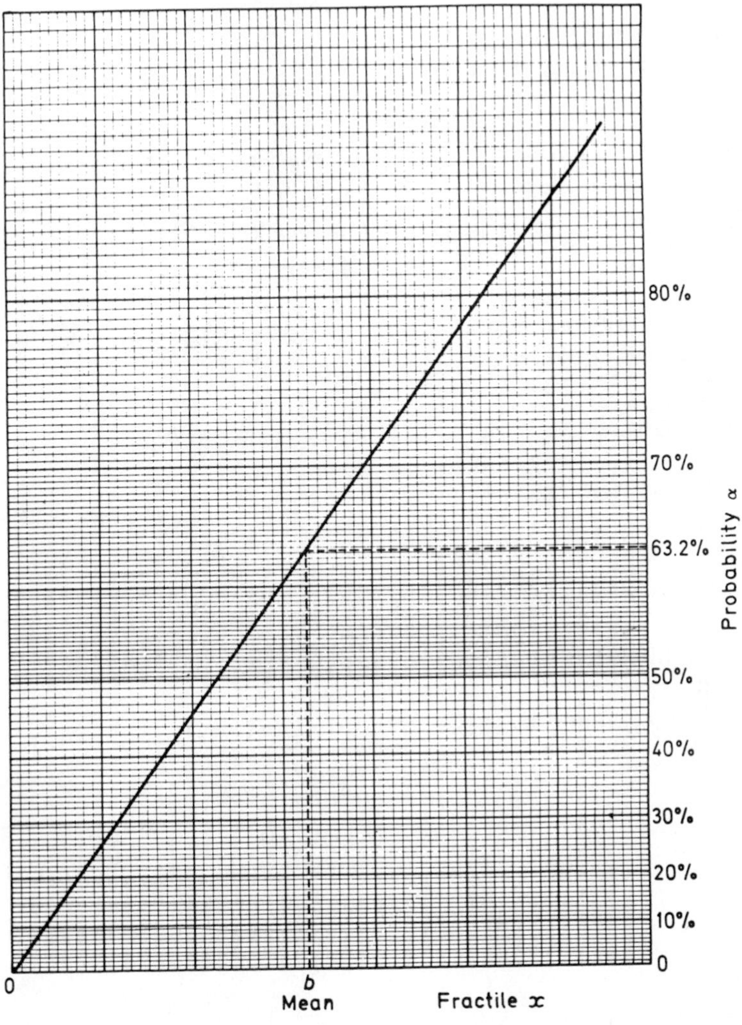

Figure 11.4 Exponential distribution, probability paper

12 EXTREME VALUE

We consider the distribution of the smallest extreme. Reversal of the sign of x gives the distribution of the largest extreme.

Range $-\infty \leqslant x \leqslant +\infty$
Location parameter a, the mode
Scale parameter $b > 0$

Distribution function	$1 - \exp\{-\exp[(x-a)/b]\}$
Probability density function	$(1/b)\exp[(x-a)/b] \times \exp\{-\exp[(x-a)/b]\}$
Inverse distribution function (of probability α)	$a + b \log\log[1/(1-\alpha)]$
Survival function	$\exp\{-\exp[(x-a)/b]\}$
Inverse survival function (of probability α)	$a + b \log\log(1/\alpha)$
Hazard function	$(1/b)\exp[(x-a)/b]$
Cumulative hazard function	$\exp[(x-a)/b]$
Moment generating function	$\exp(at/b)\,\Gamma(1+t)$
Mean	$a + b\Gamma'(1)$ $\Gamma'(1) = -0.57721$ is the first derivative of the gamma function $\Gamma(n)$ with respect to n at $n = 1$
Variance	$b^2\pi^2/6$
Standard deviation	$b\pi/6^{1/2}$
Mode	a
Median	$a + b \log\log 2$

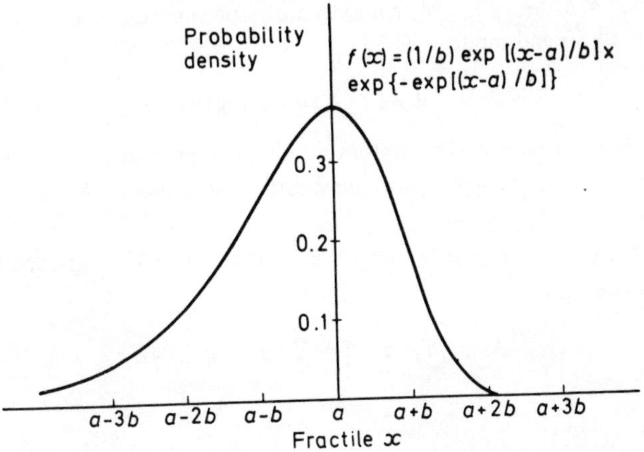

Figure 12.1 Probability density function for the extreme value variate **X** *: a, b (smallest extreme)*

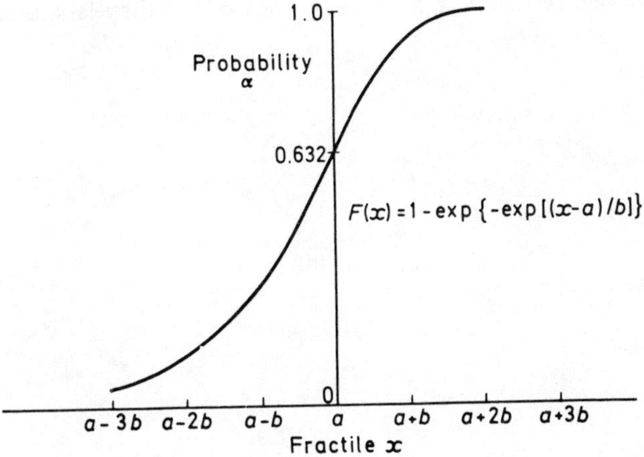

Figure 12.2 Distribution function for the extreme value variate **X** *: a, b (smallest extreme)*

12.1 Variate relationships

1. Let $\mathbf{X}_i : a, b; i = 1, \ldots, n$ be n extreme value variates, each with scale parameter b and mode a. Then

$$\text{Min}_i [\mathbf{X}_i : a, b] \sim \mathbf{X} : a - b \log (n), b$$

   That is, the smallest of n random numbers of an extreme value variate is distributed as an extreme value variate with mode $a - b \log n$.

2. For the distribution of the largest extreme, y:a, b, the corresponding relationship is

$$\text{Max}_i [\mathbf{Y}_i : a, b] \sim \mathbf{Y} : a + b \log (n), b$$

12.2 Random number generation

Let $\mathbf{R}$ denote a unit rectangular variate. Random numbers of the extreme value variate $\mathbf{X}$:a, b can be computed from the relationship

$$\mathbf{X} : a, b \sim a + b \log \log \mathbf{R}$$

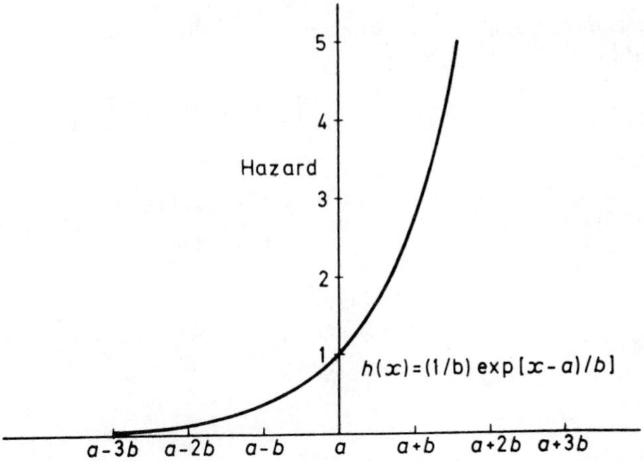

Figure 12.3 *Hazard function for the extreme value variate* $\mathbf{X}: a, b$ *(smallest extreme)*

13 F (Variance ratio)

Variate $\mathbf{F}: v, w$
Range $0 \leqslant x \leqslant +\infty$
Shape parameters v, w, positive integers, referred to as degrees of freedom

Probability density function	$\dfrac{\Gamma[\frac{1}{2}(v+w)](v/w)^{v/2}x^{(v-2)/2}}{\Gamma(\frac{1}{2}v)\Gamma(\frac{1}{2}w)(1+v/w)x^{(v+w)/2}}$
r^{th} moment about the origin	$\dfrac{(w/v)^r\Gamma(\frac{1}{2}v+r)\Gamma(\frac{1}{2}w-r)}{\Gamma(\frac{1}{2}v)\Gamma(\frac{1}{2}w)}$, $w>2r$
Mean	$w/(w-2)$, $w>2$
Variance	$2w^2(v+w-2)/v(w-2)^2(w-4)$, $w>4$
Mode	$w(v-2)/v(w+2)$, $v>1$
Coefficient of skewness	$\dfrac{(2v+w-2)[8(w-4)]^{1/2}}{(w-6)(v+w-2)^{1/2}}$, $w>6$
Coefficient of variation	$[2(v+w-2)/v(w-4)]^{1/2}$, $w>4$

13.1 Variate relationships

1. The fractile of the variate $\mathbf{F}: v, w$ at probability level $1-\alpha$ is the reciprocal of the fractile of the variate $\mathbf{F}: w, v$ at probability level α. That is

$$G_{\mathbf{F}}(1-\alpha: v, w) = 1/G_{\mathbf{F}}(\alpha: w, v)$$

where $G_{\mathbf{F}}(\alpha: v, w)$ is the inverse distribution function of $\mathbf{F}: v, w$ at probability level α.

2. The variate $\mathbf{F}: v, w$ is related to the chi-squared variates $\boldsymbol{\chi}^2: v$ and $\boldsymbol{\chi}^2: w$ by

$$\mathbf{F}: v, w \sim (w\boldsymbol{\chi}^2: v)/(v\boldsymbol{\chi}^2: w)$$

3. As the degrees of freedom v and w increase the $\mathbf{F}$ variate tends to normality.

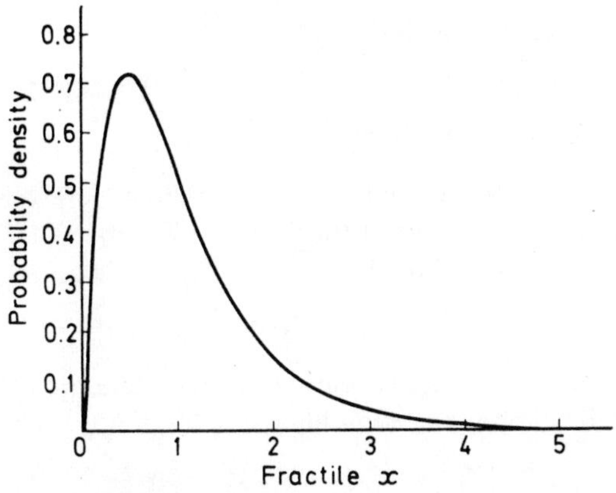

Figure 13.1 Probability density function for the **F** *variate* **F**: *4, 40*

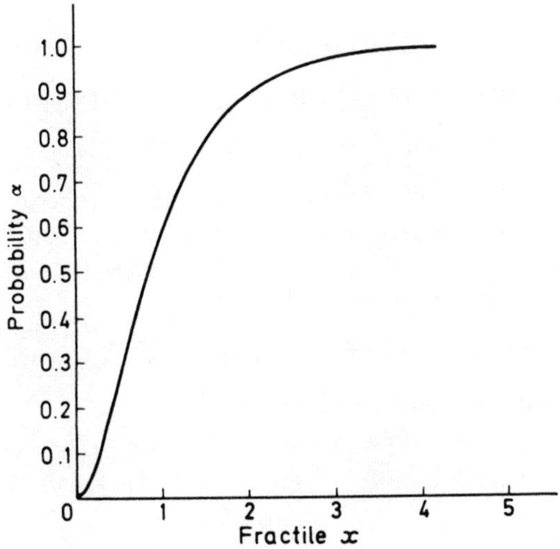

Figure 13.2 Distribution function for the **F** *variate* **F**: *4, 40*

4. The variate $\mathbf{F}: v, \infty$ is related to the chi-squared variate $\boldsymbol{\chi}^2 : v$ by the following equivalent statements

$$\mathbf{F}: v, \infty \sim (1/v)(\boldsymbol{\chi}^2 : v)$$

$$\text{Prob}\,[(\mathbf{F}: v, \infty) \leqslant x] = \text{Prob}\,[(\boldsymbol{\chi}^2 : v) \leqslant vx] = \alpha$$

$$G_{\mathbf{F}}(\alpha : v, \infty) = (1/v)\, G_{\boldsymbol{\chi}^2}(\alpha : v)$$

5. The fractile of the variate $\mathbf{F}: 1, w$ at probability level α is equal to the square of the fractile of the Student's $\mathbf{T}$ variate $\mathbf{T}: w$ at probability level $\frac{1}{2}(1 + \alpha)$. That is

$$G_{\mathbf{F}}(\alpha : 1, w) = [G_{\mathbf{T}}(\tfrac{1}{2}(1 + \alpha) : w)]^2$$

where G is the inverse distribution function. In terms of the inverse survival function the relationship is

$$Z_{\mathbf{F}}(\alpha : 1, w) = [Z_{\mathbf{T}}(\tfrac{1}{2}\alpha : w)]^2$$

Tables of the inverse survival function of the Student's $\mathbf{T}$ distribution are usually two-tailed so that the probability level shown is twice the α value.

6. The variate $\mathbf{F}: v, w$ and the beta variate $\boldsymbol{\beta}: w/2, v/2$ are related by

$$\text{Prob}\,[(\mathbf{F}: v, w) > x] = \text{Prob}\,[(\boldsymbol{\beta}: w/2, v/2) \leqslant w/(w + vx)]$$

$$= S_{\mathbf{F}}(x : v, w) = F_{\boldsymbol{\beta}}([w/(w + vx)] : w/2, v/2)$$

where S is the survival function and F is the distribution function. Hence the inverse survival function $Z_{\mathbf{F}}(\alpha : v, w)$ of the variate $\mathbf{F}: v, w$ and the inverse distribution function $G_{\boldsymbol{\beta}}(\alpha : w/2, v/2)$ of the beta variate $\boldsymbol{\beta}: w/2, v/2$ are related by

$$Z_{\mathbf{F}}(\alpha : v, w) = G_{\mathbf{F}}((1 - \alpha) : v, w) = (w/v)\,\{[1/G_{\boldsymbol{\beta}}(\alpha : w/2, v/2)] - 1\}$$

where α denotes probability.

7. Consider two sets of normal variates $\mathbf{N}_{1i} : \mu_1, \sigma_1; i = 1, \ldots, n_1$ and $\mathbf{N}_{2j} : \mu_2, \sigma_2; j = 1, \ldots, n_2$. Define variates $\bar{\mathbf{x}}_1, \bar{\mathbf{x}}_2, \mathbf{s}_1^2, \mathbf{s}_2^2$ as follows

$$\bar{\mathbf{x}}_1 \sim \sum_{i=1}^{n_1} (\mathbf{N}_{1i} : \mu_1, \sigma_1)/n_1$$

$$\bar{\mathbf{x}}_2 \sim \sum_{j=1}^{n_2} (\mathbf{N}_{2j} : \mu_2, \sigma_2)/n_2$$

$$s_1^2 \sim \sum_{i=1}^{n_1} [(\mathbf{N}_{1i} : \mu_1, \sigma_1) - \bar{x}_1]^2/n_1$$

$$s_2^2 \sim \sum_{j=1}^{n_2} [(\mathbf{N}_{2j} : \mu_2, \sigma_2) - \bar{x}_2]^2/n_2$$

Then

$$\mathbf{F} : n_1, n_2 \sim \left[\frac{n_1\, s_1^2}{(n_1 - 1)\sigma_1^2} \Big/ \frac{n_2\, s_2^2}{(n_2 - 1)\,\sigma_2^2}\right]$$

8. The variate $\mathbf{F} : v, w$ is related to the binomial variate with Bernoulli trial parameter $\frac{1}{2}(w + v - 2)$ and Bernoulli probability parameter p by

$$\text{Prob}\{\mathbf{F} : v, w < wp/[v(1-p)]\}$$
$$= 1 - \text{Prob}\,[\mathbf{B} : \tfrac{1}{2}(w + v - 2),\ p \leqslant \tfrac{1}{2}(v - 2)]$$

where $w + v$ is an even integer.

The case where the shape parameter is an integer is also treated separately as the Erlang distribution.

Variate $\gamma: b, c$
Range $0 \leqslant x \leqslant +\infty$
Scale parameter $b > 0$. Alternative parameter λ, $\lambda = 1/b$
Shape parameter $c > 0$
Distribution function. See Erlang distribution for case where c is an integer

Probability density function	$(x/b)^{c-1} [\exp(-x/b)] / b\Gamma(c)$ where $\Gamma(c)$ is the gamma function with parameter c, given by $\Gamma(c) = \int_0^\infty \exp(-u)\, u^{c-1}\, du$ Some properties of gamma functions are given in Section 5.2.
Moment generating function	$(1 - bt)^{-c}, t > 1/b$
Laplace transform of the p.d.f.	$(1 + bs)^{-c}$
Characteristic function	$(1 - ibt)^{-c}$
Cumulant function	$-c \log(1 - ibt)$
r^{th} cumulant	$(r-1)!\, cb^r$
r^{th} moment about the origin	$b^r \prod_{i=0}^{r-1} (c + i)$
Mean	bc
Variance	$b^2 c$
Standard deviation	$bc^{1/2}$
Mode	$b(c-1),\ c \geqslant 1$
Coefficient of skewness	$2c^{-1/2}$
Coefficient of kurtosis	$3 + 6/c$
Coefficient of variation	$c^{-1/2}$

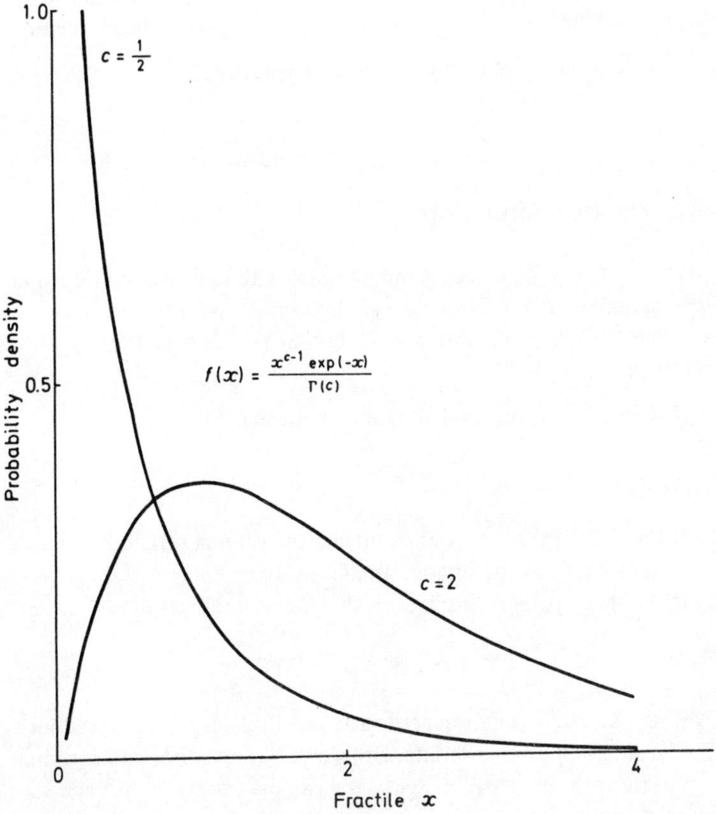

Figure 14.1 Probability density function for the gamma variate γ*: 1, c*

GAMMA

14.1 Parameter estimation

Parameter	Estimator	Method
Scale parameter, b	$s^2/\bar{x}$	Matching moments
Shape parameter, c	$(\bar{x}/s)^2$	Matching moments

$\bar{x}$ = sample mean; s^2 = sample variance (unadjusted)

14.2 Variate relationships

Notation: $\boldsymbol{\gamma}: b, c$ denotes a gamma variate with scale parameter b and shape parameter c. $\boldsymbol{\gamma}: c$ denotes a gamma variate with unit scale parameter and shape parameter c. Note that $\boldsymbol{\gamma}: b, c \sim b(\boldsymbol{\gamma}: 1, c) \sim b(\boldsymbol{\gamma}: c)$.

1. If $\mathbf{E}: b$ is an exponential variate with mean b, then

$$\boldsymbol{\gamma}: b, 1 \sim \mathbf{E}: b$$

2. If the shape parameter c is an integer the gamma variate $\boldsymbol{\gamma}: c$ may be referred to as the Erlang variate.
3. If the shape parameter c is such that $2c$ is an integer then

$$\boldsymbol{\gamma}: c \sim \tfrac{1}{2}\boldsymbol{\chi}^2: 2c \sim \tfrac{1}{2}\boldsymbol{\gamma}: 2, c$$

   where $\boldsymbol{\chi}^2: 2c$ is a chi-squared variate with $2c$ degrees of freedom.
4. The sum of a gamma variate with shape parameter c_1 and a gamma variate with shape parameter c_2 is a gamma variate with shape parameter $c_1 + c_2$.

$$\boldsymbol{\gamma}: c_1 + \boldsymbol{\gamma}: c_2 \sim \boldsymbol{\gamma}: (c_1 + c_2)$$

5. The gamma variate with shape parameter c_1 and the gamma variate with shape parameter c_2 are related to the beta variate with shape parameters c_1, c_2, denoted $\beta: c_1, c_2$ by

$$(\gamma: c_1)/(\gamma: c_1 + \gamma: c_2) \sim \beta: c_1, c_2$$

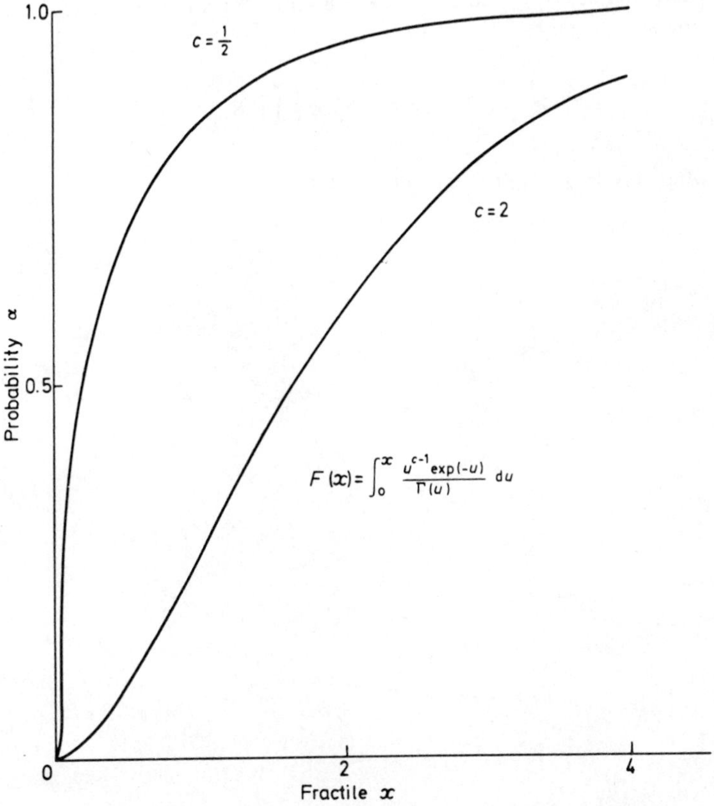

Figure 14.2 Distribution function for the gamma variate γ*: 1, c*

14.3 Random number generation

$\gamma : b, c$ denotes a gamma variate with scale parameter b and shape parameter c. For the case where c is an integer (equivalent to the Erlang variate)

$$\gamma : b, c \sim -b \log \left(\prod_{i=1}^{c} \mathbf{R}_i \right)$$

where the $\mathbf{R}_i$ are unit rectangular variates.

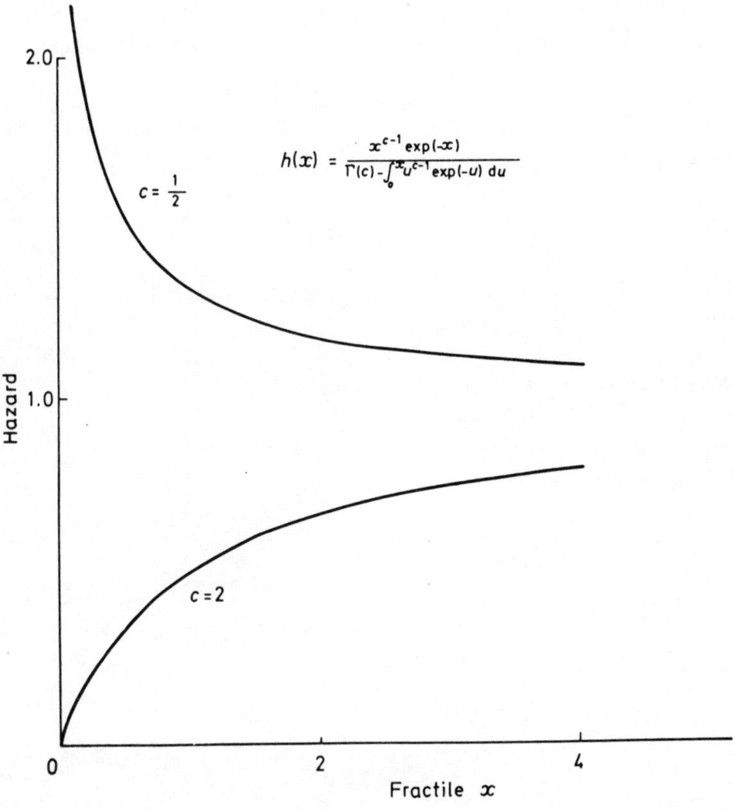

Figure 14.3 Hazard function for the gamma variate γ*: 1, c*

15 GEOMETRIC

Variate $\mathbf{G}:p$
Fractile n, number of trials
Range $n \geqslant 1$, n an integer
Given a sequence of Bernoulli trials where the probability of success at each trial is p, the geometric variate $\mathbf{G}:p$ is the number of trials up to and including the first success. Let $q = 1 - p$.
Parameter p, the Bernoulli probability parameter, $0 \leqslant p \leqslant 1$

Distribution function	$1 - q^n$
Probability function	pq^{n-1}
Inverse distribution function (of probability α)	$\log(1-\alpha)/\log(q)$
Survival function	q^n
Inverse survival function (of probability α)	$\log(\alpha)/\log(q)$

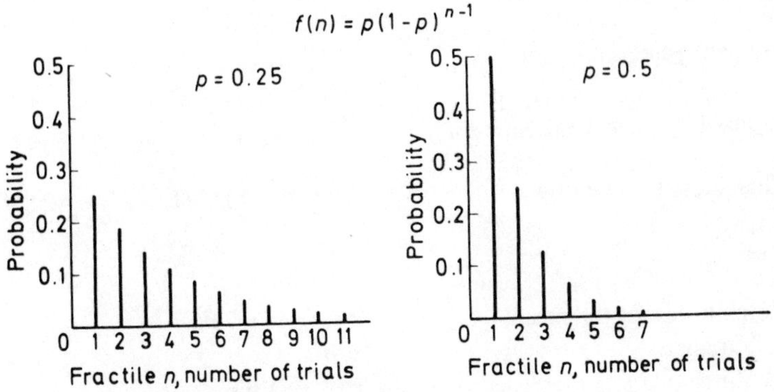

Figure 15.1 Probability function for the geometric variate G: *p*

GEOMETRIC

Hazard function $h(n) = f(n+1)/(1-F(n))$	p
Moment generating function	$p/[\exp(-t)-q]$
Probability generating function	$pt/(1-qt)$
Characteristic function	$p/[\exp(-it)-q]$
Mean	$1/p$
Moments about the mean:	
Variance	q/p^2
Third	$q(2-p)/p^3$
Fourth	$(9q^2/p^4)+(q/p^2)$
Standard Deviation	$q^{1/2}/p$
Mode	1
Coefficient of skewness	$(2-p)/q^{1/2}$
Coefficient of kurtosis	$9+p^2/q$
Coefficient of variation	$q^{1/2}$

15.1 Parameter estimation

Parameter	*Estimator*	*Property*
Bernoulli probability, p	$1/n$	Maximum likelihood

The estimator just given is applicable to a single sequence of Bernoulli trials terminating at the first success. Where x such sequences are observed the estimation of p is obtained from the Pascal variate $\mathbf{C}: x, p$.

15.2 Variate relationships

1. The relationship of the geometric to the binomial, negative binomial and Pascal variates is detailed in the section headed Bernoulli distribution.

15.3 Random number generation

Random numbers of the geometric variate $\mathbf{G}: p$ can be computed from random numbers of the unit rectangular variate $\mathbf{R}$ using the relationship

$$\mathbf{G}: p \sim \log(\mathbf{R})/\log(1-p) \text{ rounded up to the next larger integer}$$

16 HYPERGEOMETRIC

Variate $\mathbf{H}: N, X, n$
Fractile x, number of successes
Range $0 \leqslant x \leqslant \text{Min}\,[X, n]$
Suppose that from a population of N elements of which X are successes (i.e. possess a certain attribute) we draw a sample of n items without replacement. The number of successes in such a sample is a hypergeometric variate $\mathbf{H}: N, X, n$.
Parameters N, the number of elements in the population
X, the number of successes in the population
n, sample size

Probability function (probability of exactly x successes)	$\binom{X}{n}\binom{N-X}{n-x}\Big/\binom{N}{n}$
Mean	nX/N
Moments about the mean:	
Variance	$(nX/N)(1 - X/N)(N - n)/(N - 1)$
Third	$(nX/N)(1 - X/N)(1 - 2X/N)(N - n)(N - 2n)/(N - 1)(N - 2)$
Fourth	$(nX/N)(1 - X/N)(N - n)\{N(N + 1) - 6n(N - n) + (3X/N)(1 - X/N)[n(N - n)(N + 6) - 2N^2]\}/(N - 1)(N - 2)(N - 3)$
Coefficient of skewness	$(N - 2X)(N - 1)^{1/2}(N - 2n)/[nX(N - X)(N - n)(N - 2)]^{1/2}$
Coefficient of kurtosis	$[N^2(N - 1)/n(N - 2)(N - 3)(N - n)](\{[N(N + 1) - 6N(N - n)]/X(N - X)\} + [3n(N - n)(N + 6)/N^2] - 6)$
Coefficient of variation	$[(N - X)(N - n)/nX(N - 1)]^{1/2}$

16.1 Parameter estimation

To estimate the size N of the population given the number of successes in the population X, the sample size n and the number of successes in the sample.

Parameter	*Estimator*	*Property*
Population, N	nX/x	Maximum likelihood
$1/N$	x/nX	Unbiassed

16.2 Variate relationship

1. The hypergeometric variate $\mathbf{H}: N, X, n$ can be approximated by the binomial variate with Bernoulli probability parameter $p = X/N$ and Bernoulli trial parameter n, denoted $\mathbf{B}: n, p$, provided $n/N < 0.1$. That is, when the sample size is relatively small the effect of non-replacement is slight.

16.3 Random number generation

To compute random numbers of the hypergeometric variate $\mathbf{H}: N, X, n$, select n unit rectangular random numbers $\mathbf{R}_i$, $i = 1, \ldots, n$, in sequence. If $\mathbf{R}_i < p_i$ record a success, where

$$\begin{aligned} p_1 &= X/N \\ N_1 &= N \\ p_{i+1} &= (N_i p_i - d)(N - i) \\ d &= 1 \text{ if } \mathbf{R}_i \geqslant p_i \\ d &= 0 \text{ if } \mathbf{R}_i < p_i \end{aligned}$$

16.4 Note

Successive values of the probability function, $f(x)$, are related by

$$f(x+1) = f(x)(n-x)(X-x)/(x+1)(N-n-X+x+1)$$

17 LOGISTIC

Range $-\infty \leqslant x \leqslant +\infty$
Location parameter a, the mean
Scale parameter b, the standard deviation
Alternative scale parameter $k = 3^{1/2} b/\pi > 0$

Distribution function	$1 - \{1 + \exp[(x-a)/k]\}^{-1}$ $= \{1 + \exp[-(x-a)/k]\}^{-1}$ $= \frac{1}{2}\{1 + \tanh[\frac{1}{2}(x-a)/k]\}$
Probability density function	$\dfrac{\exp[(x-a)/k]}{k\{1+\exp[(x-a)/k]\}^2}$ $= \dfrac{\operatorname{sech}^2[(x-a)/2k]}{4k}$
Inverse distribution function (of probability α)	$a + k\log[\alpha/(1-\alpha)]$
Survival function	$\{1 + \exp[(x-a)/k]\}^{-1}$
Inverse survival function (of probability α)	$a + k\log[(1-\alpha)/\alpha]$
Hazard function	$\dfrac{\exp[(x-a)/k]}{k\{1+\exp[(x-a)/k]\}}$
Cumulative hazard function	$\log\{1 + \exp[(x-a)/k]\}$
Moment generating function	$\exp(at)\Gamma(1-kt)\Gamma(1+kt)$ $= \pi kt \operatorname{cosech}(\pi kt)$
Characteristic function	$\exp(\mathrm{i}at)\pi kt \operatorname{cosech}(\pi kt)$
Mean	a
Variance	$b^2 = k^2\pi^2/3$
Standard deviation	$b = k\pi/3^{1/2}$
Mode	a
Median	a
Coefficient of skewness	0

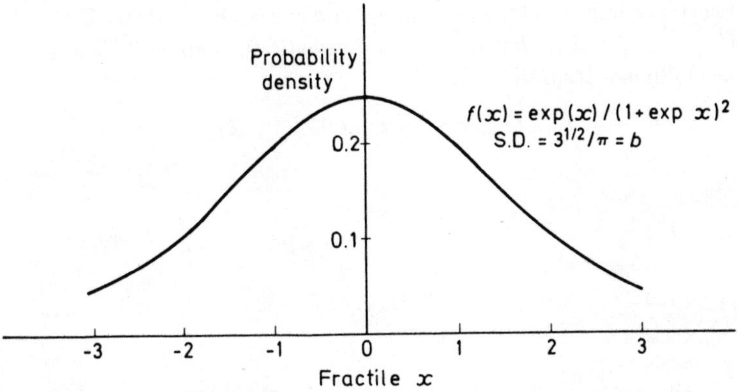

Figure 17.1 Probability density function for the logistic variate with a = 0, k = 1

Coefficient of kurtosis	4.2
Coefficient of variation	$\pi k/a3^{1/2}$

17.1 Random number generation

Let **R** denote a unit rectangular variate. Random numbers of the logistic variate with location parameter a and scale parameter k ($k = 3^{1/2}\, b/\pi$ where b is the standard deviation), denoted $\mathbf{X}: a, k$, can be computed from the relation

$$\mathbf{X}: a, k \sim a + k \log [\mathbf{R}/(1 - \mathbf{R})]$$

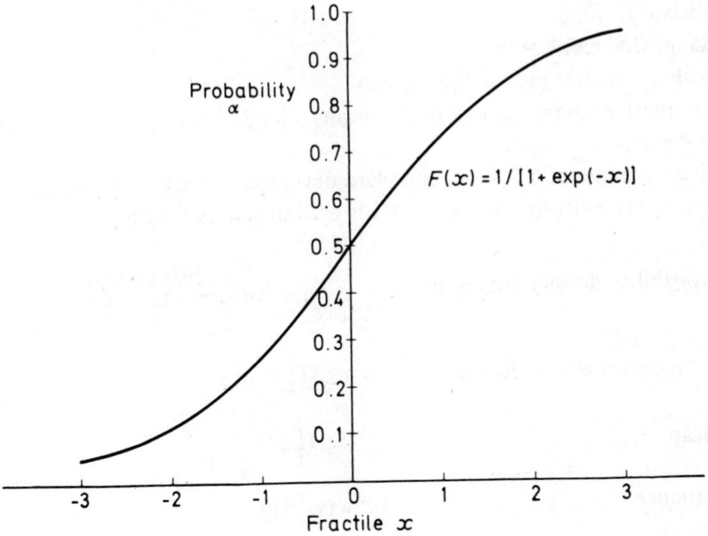

Figure 17.2 Distribution function for the logistic variate with a = 0, k = 1

18 LOGNORMAL

Variate **L** : m, σ

Range $0 \leqslant x \leqslant +\infty$

Scale parameter $m > 0$, the median

Alternative parameter $\mu > 0$, the mean of log **L**; m and μ are related by $m = \exp \mu$, $\mu = \log m$.

Shape parameter $\sigma > 0$, the standard deviation of log **L**. For compactness the substitution $w = \exp(\sigma^2)$ is used in several formulae.

Probability density function	$\dfrac{1}{x\sigma(2\pi)^{1/2}} \exp\left\{\dfrac{-[\log(x/m)]^2}{2\sigma^2}\right\}$
r^{th} moment about the origin	$m^r \exp(\frac{1}{2} r^2 \sigma^2)$
Mean	$m \exp(\frac{1}{2}\sigma^2)$
Variance	$m^2 w(w-1)$
Standard deviation	$m(w^2 - w)^{1/2}$
Mode	m/w
Median	m
Coefficient of skewness	$(w+2)(w-1)^{1/2}$
Coefficient of kurtosis	$w^4 + 2w^3 + 3w^2 - 3$
Coefficient of variation	$(w-1)^{1/2}$

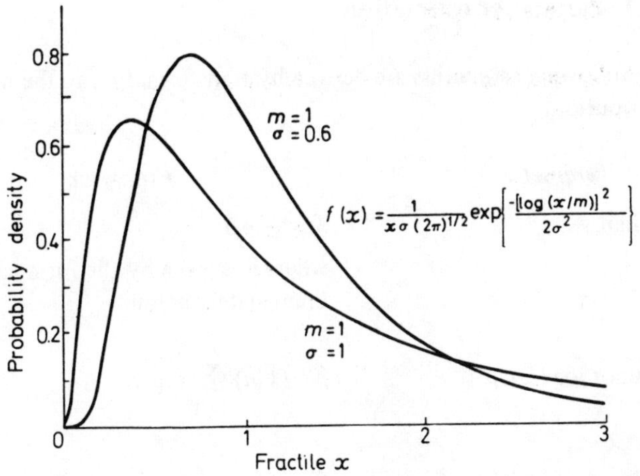

Figure 18.1 Probability density function for the lognormal variate L: *m, σ*

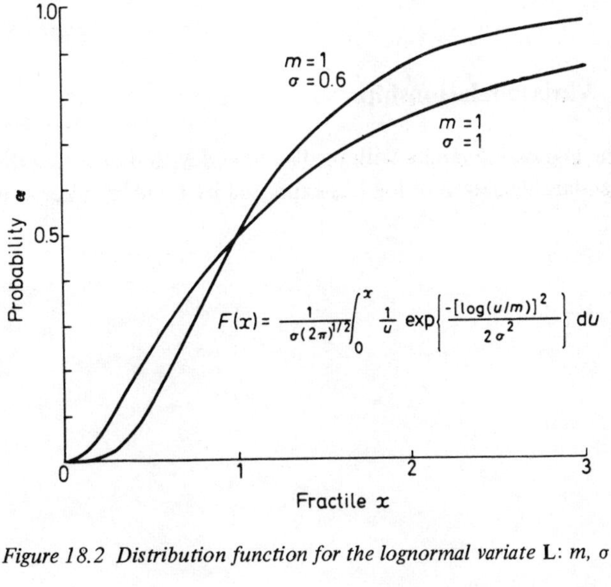

Figure 18.2 Distribution function for the lognormal variate L: *m, σ*

18.1 Parameter estimation

The following estimators are derived by transformation to the normal distribution.

Parameter	*Estimator*
Median, m	$\hat{m} = \exp \hat{\mu}$ where $\hat{\mu}$ is given by the formula immediately below
Mean of log (**L**), μ	$\hat{\mu} = (1/n) \sum_{i=1}^{n} \log x_i$
Variance of log (**L**), σ^2	$\hat{\sigma}^2 = [1/(n-1)] \sum_{i=1}^{n} [\log (x_i - \hat{\mu})]^2$

Observations $x_i, i = 1, \ldots, n$

18.2 Variate relationships

1. The lognormal variate with median m and with σ denoting the standard deviation of log **L** is expressed by **L** : m, σ. Alternatively

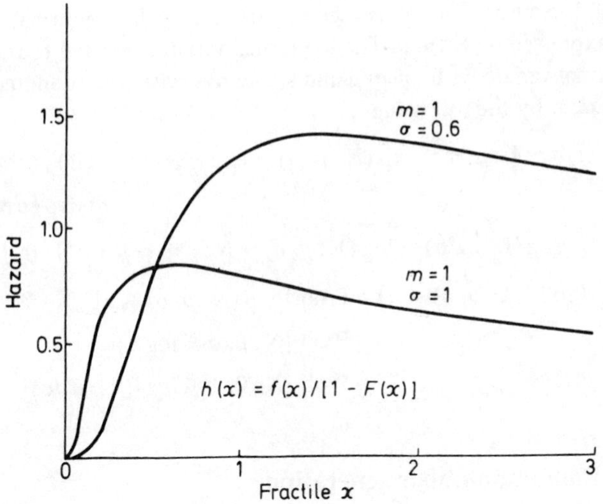

Figure 18.3 Hazard function for the lognormal variate L: m, σ

if μ, the mean of log **L**, is used as a parameter the lognormal variate is expressed by $\mathbf{L}: \mu, \sigma$. The lognormal variate is related to the normal variate with mean μ and standard deviation σ, denoted $\mathbf{N}: \mu, \sigma$, by the following

$$\mathbf{L}: m, \sigma \sim \mathbf{L}: \mu, \sigma \sim \exp(\mathbf{N}: \mu, \sigma) \sim \exp(\mu + \sigma\mathbf{N}: 0, 1)$$

$$\sim m \exp(\sigma\mathbf{N}: 0, 1)$$

$$\log(\mathbf{L}: m, \sigma) \sim \log(\mathbf{L}: \mu, \sigma) \sim \mathbf{N}: \mu, \sigma \sim \mu + \sigma\mathbf{N}: 0, 1$$

$$\begin{aligned} \text{Prob}\ [(\mathbf{L}: \mu, \sigma) \leqslant x] &= \text{Prob}\ [\exp(\mathbf{N}: \mu, \sigma) \leqslant x] \\ &= \text{Prob}\ [\mathbf{N}: \mu, \sigma \leqslant \log x] \\ &= \text{Prob}\ [\mathbf{N}: 0, 1 \leqslant \log((x - \mu)/\sigma)] \end{aligned}$$

18.3 Random number generation

The relationship of the lognormal variate with median m and shape parameter σ, denoted $\mathbf{L}: m, \sigma$, to the unit normal variate $\mathbf{N}: 0, 1$ and the relationship of the latter to the unit rectangular variate **R** give

$$\mathbf{L}: m, \sigma \sim m \exp(\sigma\mathbf{N}: 0, 1)$$

$$\mathbf{L}: m, \sigma \approx m \exp\left[\sigma\left(\sum_{i=1}^{12} \mathbf{R}_i - 6\right)\right]$$

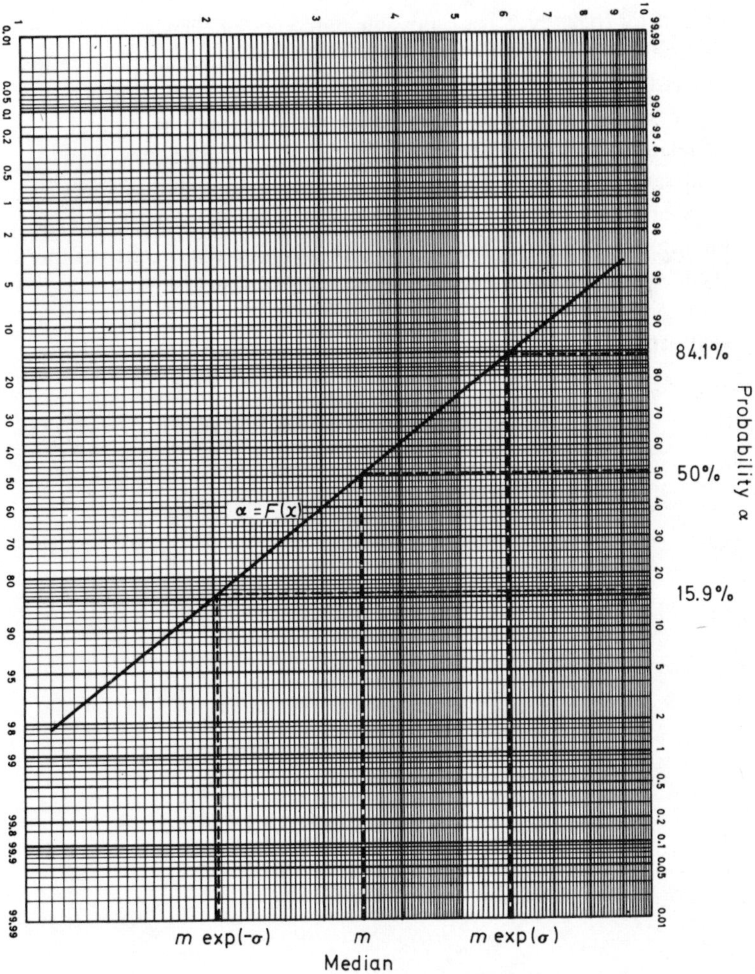

Figure 18.4 Lognormal distribution, probability paper

19 MULTINOMIAL

Consider a trial which has k possible outcomes labelled A_i, $i = 1, \ldots, k$. Outcome A_i occurs with probability p_i. The multinomial distribution relates to a set of n independent trials of this type. We define a *multivariate* as a vector each of whose elements is a variate. In general correlations may exist between these elements. The multinomial multivariate is $\mathbf{M} = [\mathbf{M}_i]$ where $\mathbf{M}_i$ is the variate 'number of times event A_i occurs'. The fractile of a multivariate is a vector $x = [x_i]$. For the multinomial variate, x_i is the fractile of $\mathbf{M}_i$ and is the number of times event A_i occurs in the n trials. The probability function $f(x_1, \ldots, x_k)$ is the probability that event A_i occurs x_i times, $i = 1, \ldots, k$, in the n trials, and is given by

$$f(x_1, \ldots, x_k) = n! \prod_{i=1}^{k} (p_i^{x_i}/x_i!)$$

20 NEGATIVE BINOMIAL

Variate $\mathbf{Y}: x, p$
Fractile y, number of failures
Range $0 \leqslant y \leqslant +\infty$, y an integer
The negative binomial variate $\mathbf{Y}: x, p$ is the number of failures before the xth success in a sequence of Bernoulli trials where the probability of success at each trial is p and the probability of failure is $q = 1 - p$.
Parameters x, the Bernoulli success parameter, a positive integer
p, the Bernoulli probability parameter, $0 < p < 1$

Distribution function	$\sum_{i=1}^{y} \binom{x+i-1}{i} p^x q^i$
Probability function	$\binom{x+y-1}{y} p^x q^y$
Moment generating function	$p^x(1 - q \exp t)^{-x}$
Probability generating function	$p^x(1 - qt)^{-x}$
Characteristic function	$p^x [1 - q \exp(\mathrm{i}t)]^{-x}$
Cumulant function	$x \log(p) - x \log(1 - q \exp t)$
Cumulants:	
First	xq/p
Second	xq/p^2
Third	$xq(1+q)/p^3$
Fourth	$xq(6q + p^2)/p^4$
Mean	xq/p
Moments about the mean:	
Variance	xq/p^2
Third	$xq(1+q)/p^3$
Fourth	$(xq/p^4)(3xq + 6q + p^2)$
Standard deviation	$(xq)^{1/2}/p$
Coefficient of skewness	$(1+q)(xq)^{-1/2}$
Coefficient of kurtosis	$3 + \frac{6}{x} + \frac{p^2}{xq}$

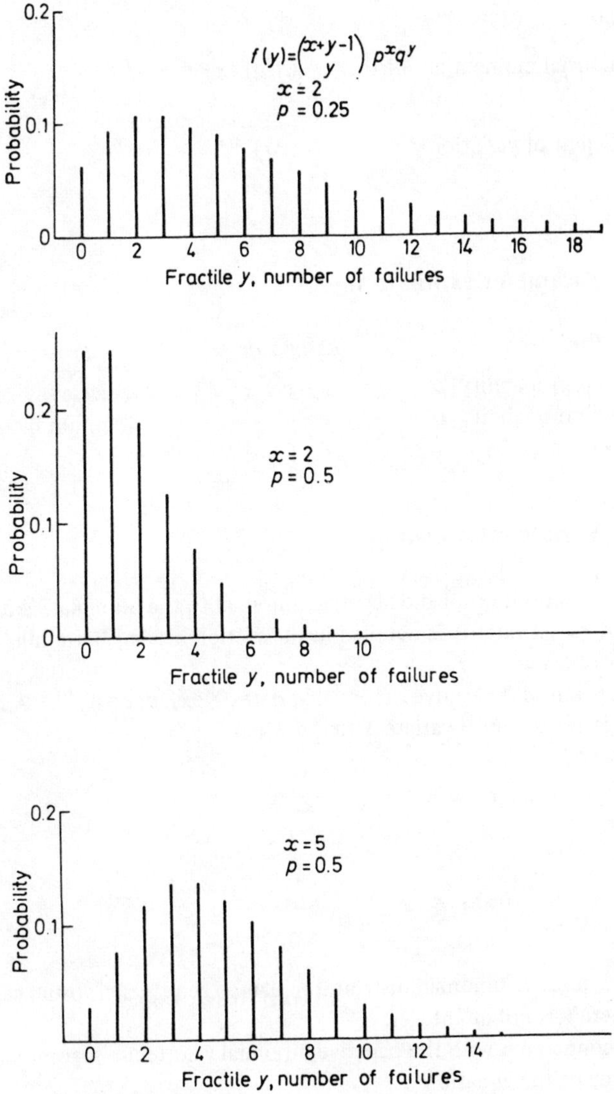

Figure 20.1 Probability function for the negative binomial variate **Y**: *x*, *p*

Factorial moment generating function	$(1 - q^t/p)^{-x}$
r^{th} factorial moment about the origin	$(q/p)^r (x + r - 1)^r$
Coefficient of variation	$(xq)^{-1/2}$

20.1 Parameter estimation

Parameter	*Estimator*	*Property*
Bernoulli probability, p	$(x - 1)/(y + x - 1)$	Unbiassed
Bernoulli probability, p	$x/(y + x)$	Maximum likelihood

20.2 Variate relationships

1. The relationship of the negative binomial to the binomial, geometric and Pascal variates is detailed in the section headed Bernoulli distribution.
2. The sum of k negative binomial variates $\mathbf{Y}: x_i, p; i = 1, \ldots, k$ is a negative binomial variate $\mathbf{Y}: x', p$ where

$$x' = \sum_{i=1}^{k} x_i$$

20.3 Notes

1. The negative binomial distribution can be generalised to the case where y is not an integer.
2. In connection with the negative binomial variate use is sometimes made of the equation

$$\binom{x + y - 1}{y} = (-1)^y \binom{-x}{y}$$

3. The following properties can be used as a guide in choosing between negative binomial, binomial and Poisson distribution models

Negative binomial	Variance > Mean
Binomial	Variance < Mean
Poisson	Variance = Mean

20.4 Random number generation

1. *Rejection technique:* Select a sequence of unit rectangular random numbers, recording the numbers of these which are greater than and less than p. When the number greater than p first reaches x the number less than p is a negative binomial random number.
2. *Geometric distribution method:* If p is small a method which may be faster than the rejection technique involves adding x geometric random numbers. The equation is

$$\mathbf{Y}:x,p \sim \left(\sum_{i=1}^{x} \mathbf{G}_i : p\right) - x$$

where $\mathbf{G}_i : p$ is a geometric variate, whose random numbers are generated using

$\mathbf{G}_i : p \sim \log(\mathbf{R}_i)/\log(1-p)$ rounded up to the next larger integer

$\mathbf{R}_i$ is a unit rectangular variate.

21 NORMAL

Variate **N** : μ, σ
Range $-\infty \leqslant x \leqslant +\infty$
Location parameter μ, the mean
Scale parameter $\sigma > 0$, the standard deviation

Probability density function	$\dfrac{1}{\sigma(2\pi)^{1/2}} \exp\left[\dfrac{-(x-\mu)^2}{2\sigma^2}\right]$
Moment generating function	$\exp(\mu t + \frac{1}{2}\sigma^2 t^2)$
Characteristic function	$\exp(\mathrm{i}\mu t - \frac{1}{2}\sigma^2 t^2)$
Cumulant function	$\mathrm{i}\mu t - \frac{1}{2}\sigma^2 t^2$
r^{th} cumulant	$\kappa_2 = \sigma^2,\ \kappa_r = 0,\ r > 2$
Mean	μ
r^{th} moment about the mean	$\mu_r = 0$ for r odd
	$\mu_r = \dfrac{\sigma^r r!}{2^{r/2}\,[(r/2)!]}$ for r even
Variance	σ^2
Standard deviation	σ
Mean deviation	$\sigma(2/\pi)^{1/2}$
Mode	μ
Median	μ
Standardised r^{th} moment about the mean	$\eta_r = 0$ for r odd
	$\eta_r = \dfrac{r!}{2^{r/2}\,[(r/2)!]}$ for r even
Coefficient of skewness	0
Coefficient of kurtosis	3
Information content	$\log_2\,[\sigma(2\pi e)^{1/2}]$

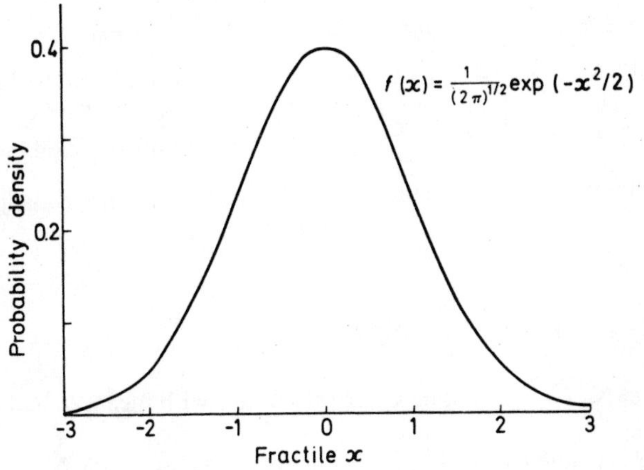

Figure 21.1 Probability density function for the standard normal variate **N**: *0, 1*

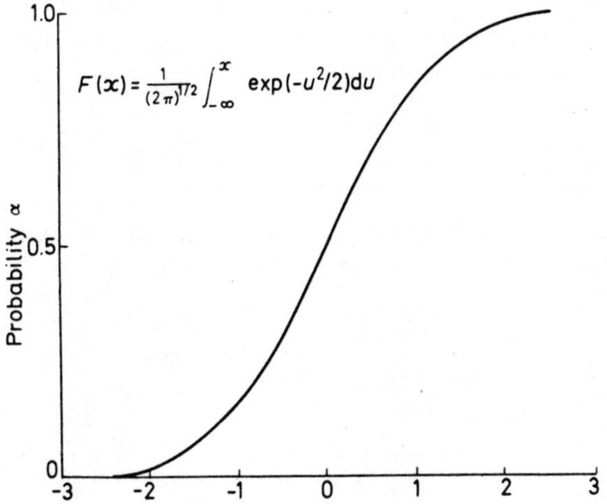

Figure 21.2 Distribution function for the standard normal variate **N**: *0, 1*

21.1 Parameter estimation

Parameter	*Estimator*	*Properties*
Mean	$\bar{x}$	Unbiassed, Maximum likelihood
Variance	$ns^2/(n-1)$	Unbiassed
Variance	s^2	Maximum likelihood

n = sample size; $\bar{x}$ = sample mean; s^2 = sample variance (unadjusted)

21.2 Variate relationships

1. Let $\mathbf{N}_i$, $i = 1, \ldots, n$ be n normal variates with means μ_i and variances σ_i^2. Then $\sum_{i=1}^{n} c_i \mathbf{N}_i$ is normally distributed with mean $\sum_{i=1}^{n} c_i\mu_i$ and variance $\sum_{i=1}^{n} c_i^2\sigma_i^2$, where the c_i, $i = 1, \ldots, n$, are weighting factors.
2. The sum of n normal variates, $\mathbf{N}: \mu, \sigma$, is a normal variate with mean $n\mu$ and standard deviation $\sigma n^{1/2}$

$$\sum_{i=1}^{n} \mathbf{N}_i : \mu, \sigma \sim \mathbf{N} : n\mu, \sigma n^{1/2}$$

3. The sum of the squares of n unit normal variates, $\mathbf{N}: 0, 1$, is a chi-squared variate with n degrees of freedom, $\boldsymbol{\chi}^2 : n$

$$\sum_{i=1}^{n} (\mathbf{N}_i : 0, 1)^2 \sim \boldsymbol{\chi}^2 : n$$

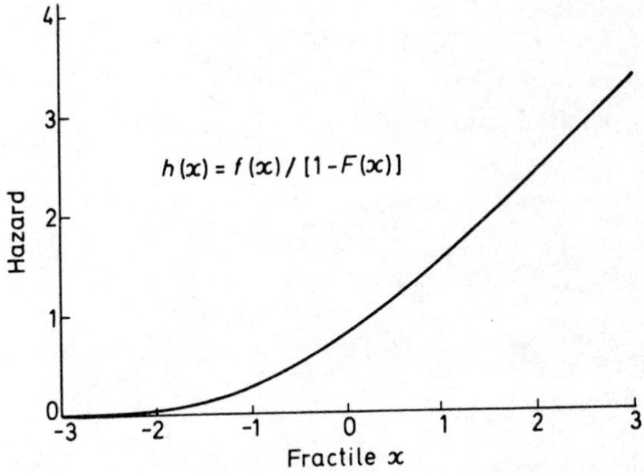

Figure 21.3 Hazard function for the standard normal variate **N**: *0, 1*

21.3 Random number generation

Let **R** denote a unit rectangular variate. The following approximation increases in accuracy with k

$$\mathbf{N}:0,1 \approx \frac{\sum_{i=1}^{k} \mathbf{R}_i - k/2}{(k/12)^{1/2}}$$

For many practical purposes it is sufficient to take $k = 12$ giving

$$\mathbf{N}:0,1 \approx \sum_{i=1}^{12} \mathbf{R}_i - 6$$

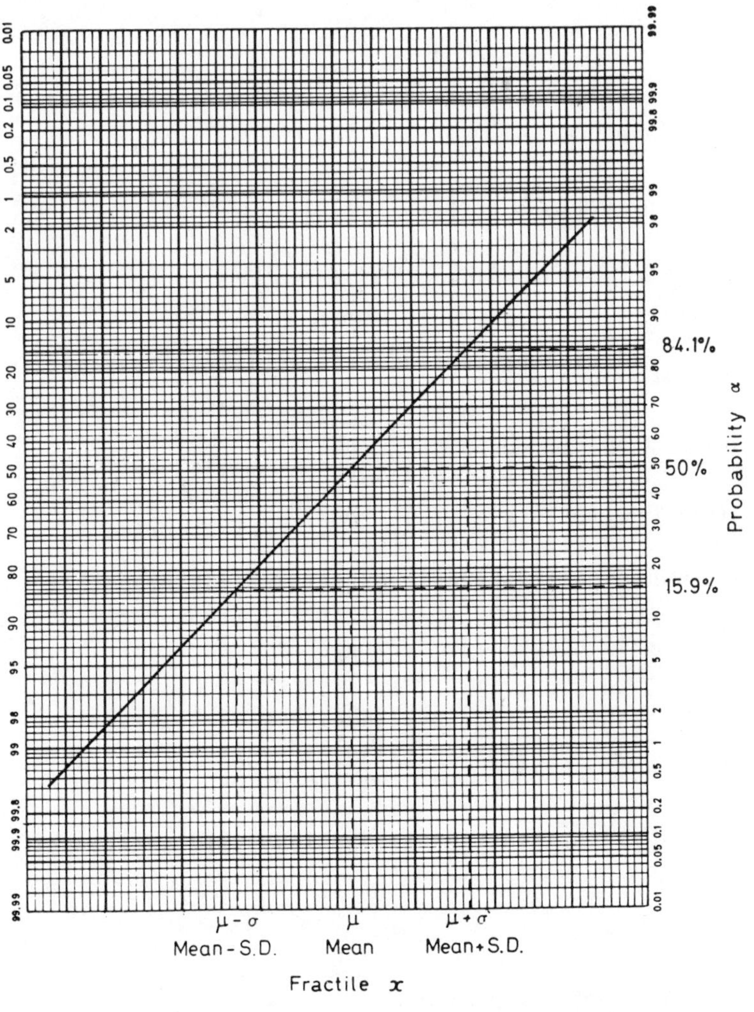

Figure 21.4 Normal distribution, probability paper

22 PARETO

Range $1 \leqslant x \leqslant +\infty$
Shape parameter $c > 0$

Distribution function	$1 - x^{-c}$
Probability density function	cx^{-c-1}
Inverse distribution function (of probability α)	$[1/(1-\alpha)]^{1/c}$
Survival function	x^{-c}
Inverse survival function	$(1/\alpha)^{1/c}$
Hazard function	c/x
Cumulative hazard function	$c \log x$
r^{th} moment about the origin	$c/(c-r), c > r$
Mean	$c/(c-1),\ c > 1$
Variance	$[c/(c-2)] - [c/(c-1)]^2,\ c > 2$
Coefficient of variation	$(c-1)/[c(c-1)]^{1/2},\ c > 2$

22.1 Parameter estimation

Parameter	*Estimator*	*Property*
$1/c$	$(1/n) \sum_{i=1}^{n} \log x_i$	Maximum likelihood

Observations x_i, $i = 1, \ldots, n$.

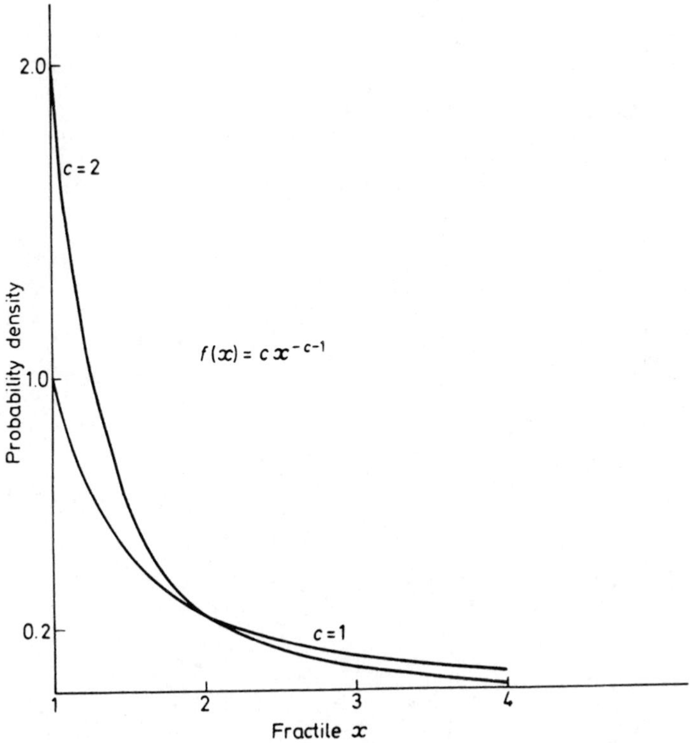

Figure 22.1 Probability density function for the Pareto variate

22.2 Variate relationship

1. The Pareto variate $\mathbf{X}:c$ is related to the unit rectangular variate $\mathbf{R}$ by

$$\mathbf{X}:c \sim (1/\mathbf{R})^{1/c}$$

22.3 Random number generation

Use variate relationship 1 immediately above.

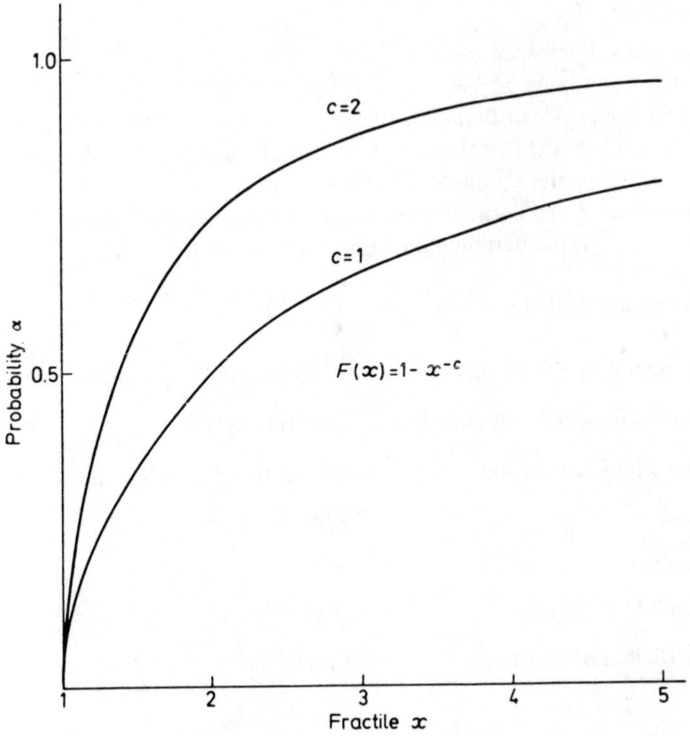

Figure 22.2 Distribution function for the Pareto variate

Variate **C** : x, p
Fractile n, number of trials
Range $n \geqslant 1$, n an integer
Given a sequence of Bernoulli trials where the probability of success at each trial is p, the Pascal variate **C** : x, p is the number of trials up to and including the xth success. Let $q = 1 - p$.
Parameters x, the Bernoulli success parameter, a positive integer
p, the Bernoulli probability parameter $0 < p < 1$

Probability function	$\binom{n-1}{n-x} p^x q^{n-x}$
Moment generating function	$p^x \exp(tx)/(1 - q \exp t)^x$
Probability generating function	$(pt)^x/(1 - qt)^x$
Characteristic function	$p^x \exp(\mathrm{i}tx)/(1 - q \exp \mathrm{i}t)^x$
Mean	x/p
Variance	xq/p^2
Standard deviation	$(xq)^{1/2}/p$
Coefficient of variation	$(q/x)^{1/2}$

23.1 Parameter estimation

Parameter	*Estimator*	*Property*
Bernoulli probability, p	$(x - 1)/(n - 1)$	Unbiassed
Bernoulli probability, p	x/n	Maximum likelihood

23.2 Variate relationships

1. The relationship of the Pascal to the binomial, negative binomial and geometric variates is detailed in the section headed Bernoulli distribution.

2. The sum of k Pascal variates $\mathbf{C}: x_i, p; i = 1, \ldots, k$ is a Pascal variate $\mathbf{C}: x', p$ where

$$x' = \sum_{i=1}^{k} x_i$$

23.3 Random number generation

Random numbers of the Pascal variate $\mathbf{C}: x, p$ can be computed by adding x geometric random numbers $\mathbf{G}_i : p; i = 1, \ldots, x$, since

$$\mathbf{C}: x, p \sim \sum_{i=1}^{x} \mathbf{G}_i : p$$

the geometric random numbers being computed from unit rectangular random numbers **R** by the relationship

$\mathbf{G}: p \sim \log(\mathbf{R})/\log(1 - p)$ rounded up to the next larger integer

Variate **P** : λ
Range $0 \leqslant x \leqslant +\infty$, x an integer
Parameter λ, the mean, $\lambda > 0$

Distribution function	$\sum_{i=0}^{x} \lambda^i \exp(-\lambda)/i!$
Probability function	$\lambda^x \exp(-\lambda)/x!$
Moment generating function	$\exp\{\lambda [\exp(t) - 1]\}$
Probability generating function	$\exp[-\lambda(1 - t)]$
Characteristic function	$\exp\{\lambda [\exp(\mathrm{i}t) - 1]\}$
Cumulant function	$\lambda [\exp(t) - 1] = \sum_{i=0}^{\infty} t^i/i!$
r^{th} cumulant	λ
Moments about the origin:	
Mean	λ
Second	$\lambda + \lambda^2$
Third	$\lambda[(\lambda + 1)^2 + \lambda]$
Fourth	$\lambda(\lambda^3 + 6\lambda^2 + 7\lambda + 1)$
r^{th} moment about the mean, μ_r	$\lambda \sum_{i=0}^{r-2} \binom{r-1}{i} \mu_i$ $r > 1$, $\mu_0 = 1$
Moments about the mean:	
Variance	λ
Third	λ
Fourth	$\lambda(1 + 3\lambda)$
Fifth	$\lambda(1 + 10\lambda)$
Sixth	$\lambda(1 + 25\lambda + 15\lambda^2)$
Standard deviation	$\lambda^{1/2}$

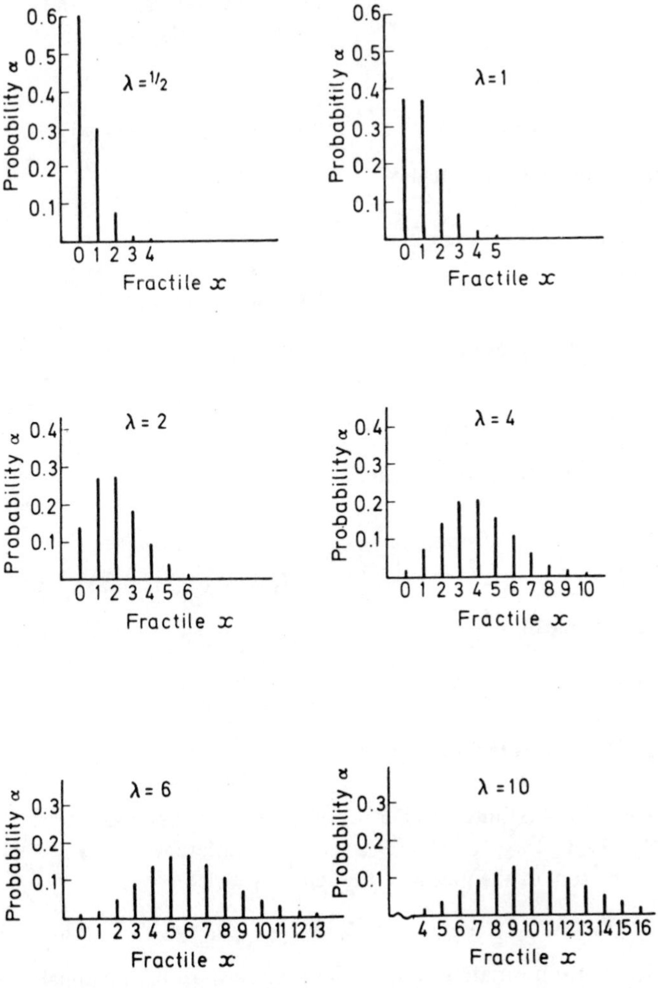

Figure 24.1 Probability function for the Poisson variate P: λ

Mode	The mode occurs when x is the largest integer less than λ. For λ an integer the values $x = \lambda$ and $x = \lambda - 1$ are tie modes.
Coefficient of skewness	$\lambda^{-1/2}$
Coefficient of kurtosis	$3 + 1/\lambda$
Factorial moments about the mean:	
Second	λ
Third	-2λ
Fourth	$3\lambda(\lambda + 2)$
Coefficient of variation	$\lambda^{-1/2}$

24.1 Parameter estimation

Parameter	*Estimator*	*Properties*
Mean, λ	$\bar{x}$	Unbiassed, Maximum likelihood

$\bar{x}$ = sample mean

24.2 Variate relationships

1. The sum of a finite number of independent Poisson variates, $\mathbf{P}_1 : \lambda_1, \mathbf{P}_2 : \lambda_2, \ldots, \mathbf{P}_n : \lambda_n$ is a Poisson variate with mean equal to the sum of the means of the separate variates

$$\mathbf{P}_1 : \lambda_1 + \mathbf{P}_2 : \lambda_2 + \ldots + \mathbf{P}_n : \lambda_n \sim \mathbf{P} : \lambda_1 + \lambda_2 + \ldots + \lambda_n$$

2. The Poisson variate $\mathbf{P} : \lambda$ is the limiting form of the binomial variate $\mathbf{B} : n, p$ as n tends to infinity, p tends to zero and np tends to λ.

$$\mathrm{Lim}_{n \to \infty,\, np \to \lambda} \left[\binom{n}{x} p^x (1 - p)^{n-x} \right] = \lambda^x \exp(-\lambda)/x!$$

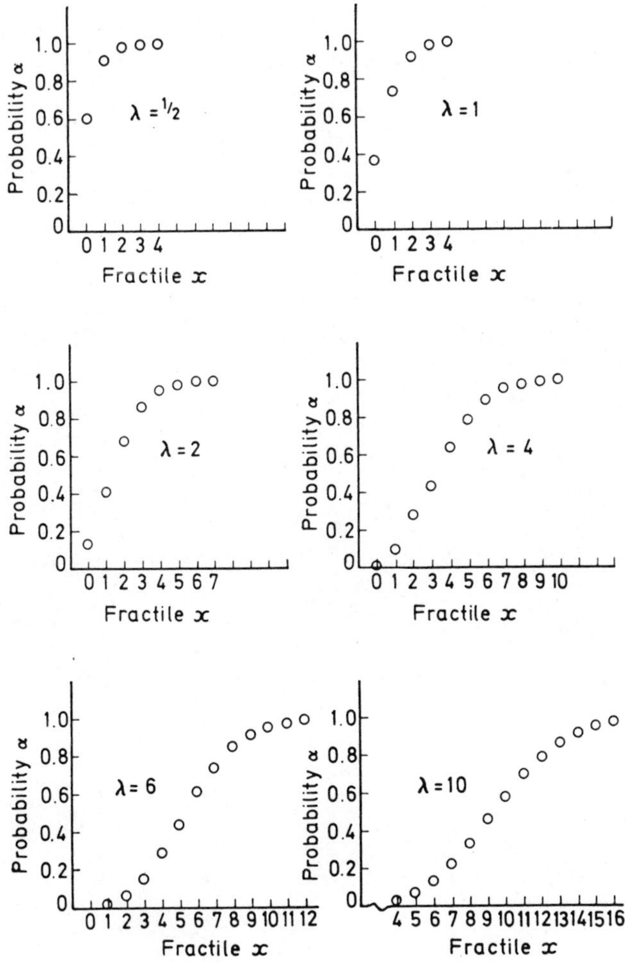

Figure 24.2 Distribution function for the Poisson variate P: λ

3. For $\lambda > 9$ the Poisson variate $\mathbf{P}:\lambda$ may be approximated by the normal variate with mean λ and variance λ.
4. The probability that the Poisson variate $\mathbf{P}:\lambda$ is less than or equal to x is equal to the probability that the chi-squared variate with $2(1+x)$ degrees of freedom, denoted $\boldsymbol{\chi}^2 : 2(1+x)$, is greater than 2λ.

$$\text{Prob}\,[\mathbf{P}:\lambda \leqslant x] = \text{Prob}\,[\boldsymbol{\chi}^2 : 2(1+x) > 2\lambda]$$

24.3 Random number generation

Calculate the distribution function $F(x)$ for $x = 0, 1, 2, \ldots, N$ where N is an arbitrary cut-off number. Choose random numbers of the unit rectangular variate, **R**. If $F(x) \leqslant \mathbf{R} < F(x+1)$ then the corresponding Poisson random number is x.

24.4 Note

Successive values of the probability function $f(x)$ are related by

$$f(x+1) = \lambda f(x)/(x+1)$$
$$f(0) = \exp(-\lambda)$$

$x = 0, 1, 2, \ldots$

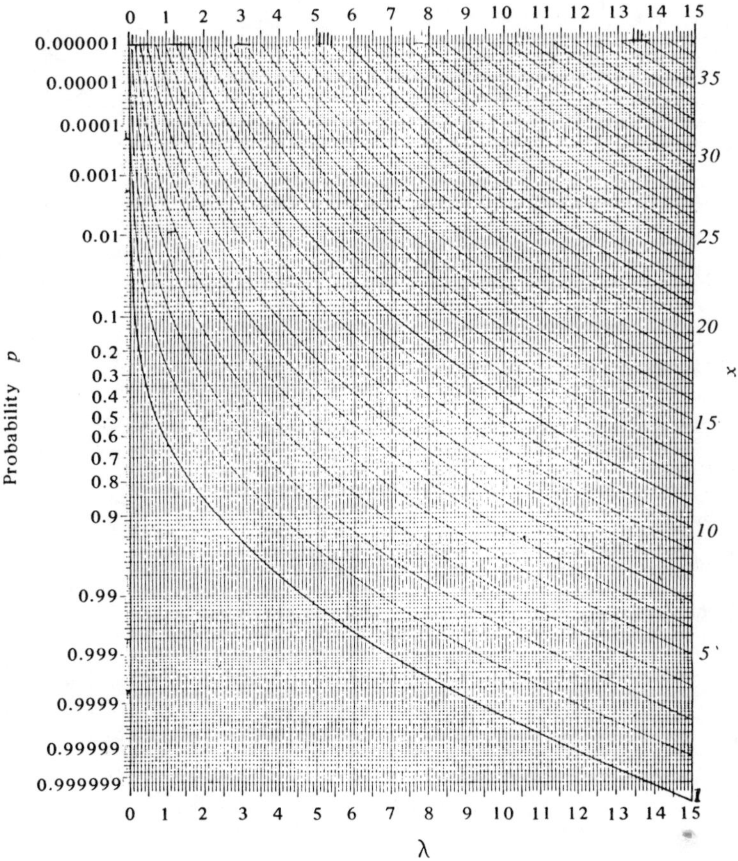

Figure 24.3 Curves for the Poisson variate **P** : λ, *showing p* = Prob [**P**: λ ⩾ *x*]

25 POWER FUNCTION

Range $0 \leqslant x \leqslant 1$
Shape parameter c

Distribution function	x^c
Probability density function	cx^{c-1}
Inverse distribution function (of probability α)	$\alpha^{1/c}$
Hazard function	$cx^{c-1}/(1 - x^c)$
Cumulative hazard function	$-\log(1 - x^c)$
r^{th} moment about the origin	$c/(c + r)$
Mean	$c/(c + 1)$
Variance	$c/(c + 2)(c + 1)^2$
Mode	1 for $c > 1$, 0 for $c < 1$
Median	$(\frac{1}{2})^{1/c}$
Coefficient of variation	$1/(c + 1)(c + 2)$

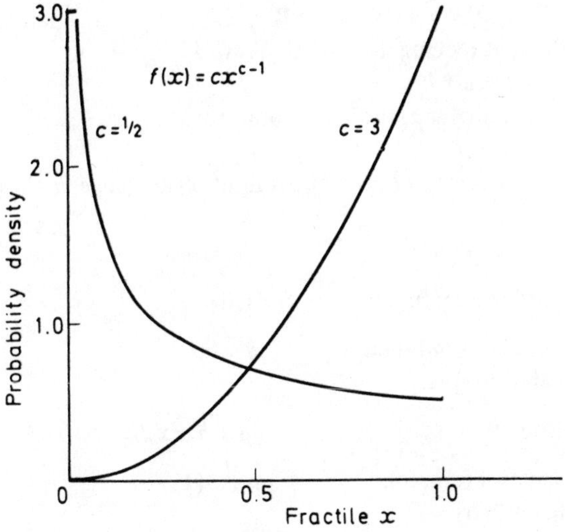

Figure 25.1 Probability density function for the power function variate

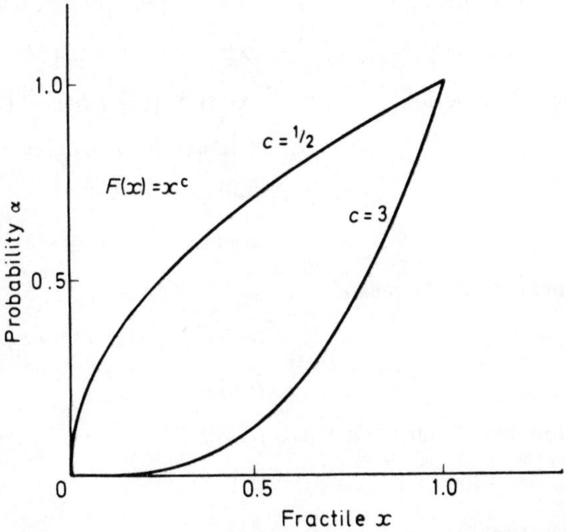

Figure 25.2 Distribution function for the power function variate

Variate **R** : a, b. Where we write **R** without specifying parameters we imply the unit rectangular variate, **R** : 0, 1.
Range $a \leqslant x \leqslant a + b$
Location parameter a, the lower limit of the range
Scale parameter b
Alternative parameter h, the upper limit of the range, $h = a + b$

Distribution function	$(x - a)/b$
Probability density function	$1/b$
Inverse distribution function (of probability α)	$a + b\alpha$
Survival function	$(a + b - x)/b$
Inverse survival function (of probability α)	$a + b(1 - \alpha)$
Hazard function	$1/(a + b - x)$
Cumulative hazard function	$\log\,[b/(a + b - x)]$
Moment generating function	$\exp\,(at)\,[\exp\,(bt) - 1]/bt$
Laplace transform of the p.d.f.	$\exp\,(-as)\,[1 - \exp\,(-bs)]/bs$
Characteristic function	$\exp\,(\mathrm{i}at)\,[\exp\,(\mathrm{i}bt) - 1]/\mathrm{i}bt$ $= \{2 \exp\,[(a + b/2)\,\mathrm{i}t] \times$ $\sinh\,(\mathrm{i}bt/2)\}/\mathrm{i}bt$
Mean	$a + b/2$
r^{th} moment about the mean	$\mu_r = 0$ for r odd $\mu_r = (b/2)^r/(r + 1)$ for r even
Variance	$b^2/12$
Fourth moment about the mean	$b^4/80$
Standard deviation	$b/12^{1/2}$
Mean deviation	$b/4$
Median	$a + b/2$

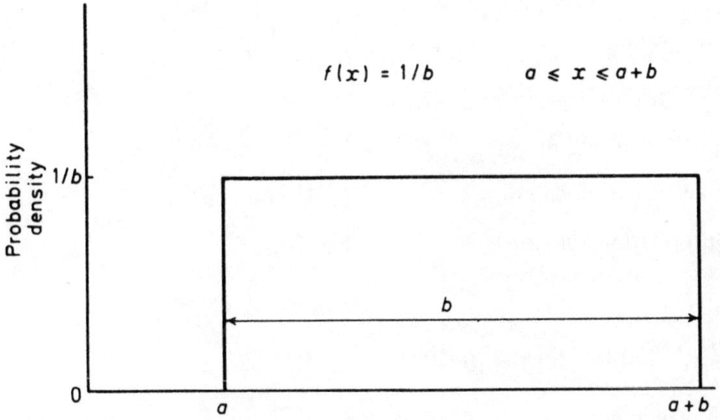

Figure 26.1 Probability density function for the rectangular variate R: *a, b*

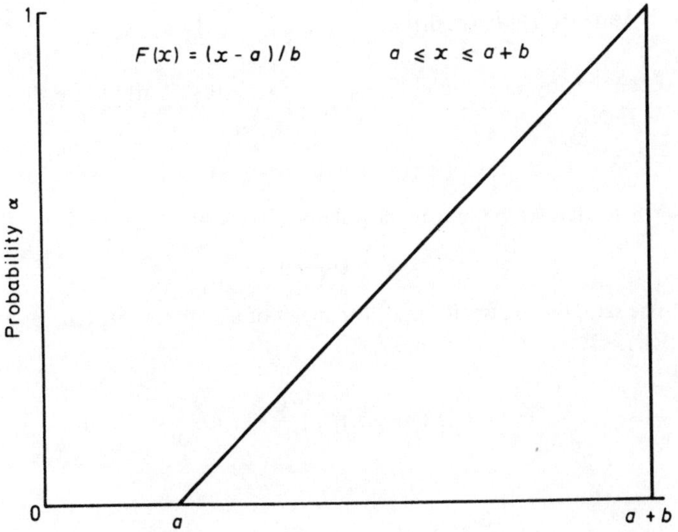

Figure 26.2 Distribution function for the rectangular variate R: *a, b*

Standardised r^{th} moment about the mean	$\eta_r = 0$ for r odd $\eta_r = 3^{r/2}/(r+1)$ for r even
Coefficient of skewness	0
Coefficient of kurtosis	9/5
Coefficient of variation	$b/[3^{1/2}(2a+b)]$
Information content	$\log_2 b$

26.1 Parameter estimation

Parameter	*Estimator*	*Method*
Lower limit, a	$\bar{x} - 3^{1/2}s$	Matching moments
Scale parameter, b	$12^{1/2}s$	Matching moments

$\bar{x}$ = sample mean; s^2 = sample variance (unadjusted)

26.2 Variate relationships

1. Let **Y** be any variate and $G_{\mathbf{Y}}$ be the inverse distribution function of **Y**, that is

$$\text{Prob}\,[\mathbf{Y} \leqslant G_{\mathbf{Y}}(\alpha)] = \alpha, \quad 0 \leqslant \alpha \leqslant 1$$

Y is related to the unit rectangular variate **R** by

$$\mathbf{Y} \sim G_{\mathbf{Y}}(R)$$

2. The sampling distribution of the mean of n unit rectangular variates $\mathbf{R}_i$, $i = 1, \ldots, n$ has the p.d.f.

$$\frac{n^n}{(n-1)!} \sum_{r=0}^{k} (-1)^r \binom{n}{r} (x - r/n)^{n-1}, \quad \frac{k}{n} \leqslant x \leqslant \frac{k+1}{n}, \; k = 0, 1, 2, \ldots, n$$

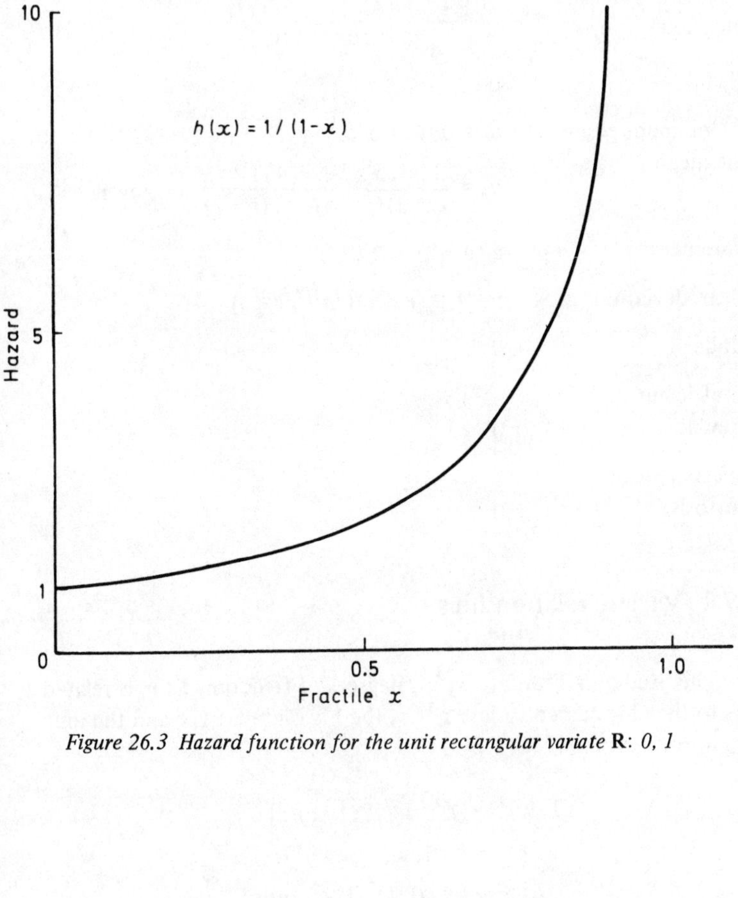

Figure 26.3 Hazard function for the unit rectangular variate **R**: *0, 1*

Variate **T** : v
Range $-\infty \leqslant x \leqslant +\infty$
Shape parameter v, degrees of freedom, v a positive integer

Probability density function	$\dfrac{\{\Gamma[(v+1)/2]\}\,[1+(x^2/v)]^{-(v+1)/2}}{(\pi v)^{1/2}\,\Gamma(v/2)}$
Mean	0
r^{th} moment about the mean	$\mu_r = 0$ for r odd $\mu_r = \dfrac{1.3.5\ldots(r-1)\,v^{r/2}}{(v-2)(v-4)\ldots(v-r)}$ for r even, $r<v$
Variance	$v/(v-2),\ v>2$
Mean deviation	$v^{1/2}\Gamma[\frac{1}{2}(v-1)]/\pi^{1/2}\Gamma(\frac{1}{2}v)$
Mode	0
Coefficient of skewness	0
Coefficient of kurtosis	0

27.1 Variate relationships

1. The Student's **T** variate with v degrees of freedom, **T** : v, is related to the chi-squared variate $\chi^2 : v$, the **F** variate **F** : 1, v and the unit normal variate **N** : 0, 1 by

$$\begin{aligned}(\mathbf{T}:v)^2 &\sim (\chi^2:1)/[(\chi^2:v)/v]\\ &\sim \mathbf{F}:1,v\\ &\sim (\mathbf{N}:0,1)^2/[(\chi^2:v)/v]\\ \mathbf{T}:v &\sim (\mathbf{N}:0,1)/[(\chi^2:v)/v]^{1/2}\end{aligned}$$

Equivalently, in terms of a probability statement

$$\text{Prob}\,[(\mathbf{T}:v)\leqslant x] = \tfrac{1}{2}\,\{1+\text{Prob}\,[(\mathbf{F}:1,v)\leqslant x^2]\}$$

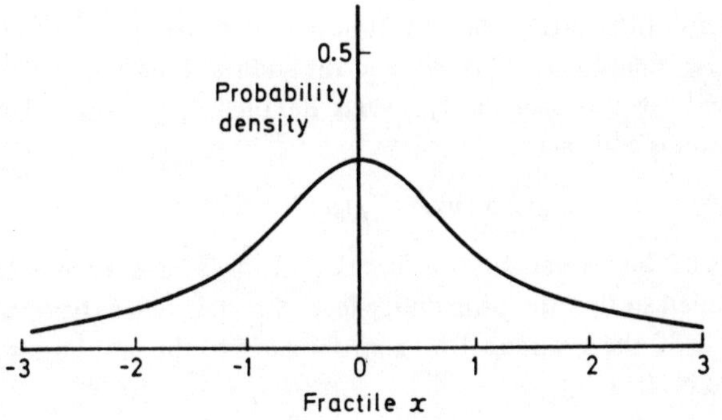

Figure 27.1 Probability density function for Student's T *variate with one degree of freedom*

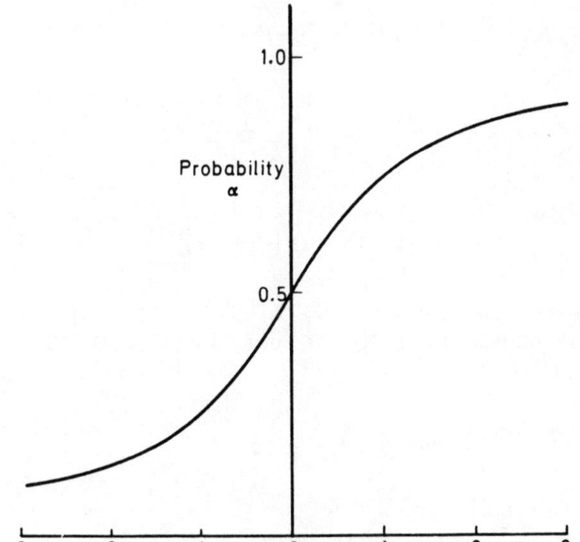

Figure 27.2 Distribution function for Student's T *variate with one degree of freedom*

In terms of the inverse survival function of $\mathbf{T}: v$ at probability level $\frac{1}{2}\alpha$, denoted $Z_{\mathbf{T}}(\frac{1}{2}\alpha : v)$, and the survival function of the $\mathbf{F}$ variate $\mathbf{F}: 1, v$ at probability level α, denoted $Z_{\mathbf{F}}(\alpha : 1, v)$, the last equation is equivalent to

$$Z_{\mathbf{T}}(\tfrac{1}{2}\alpha : v) = [Z_{\mathbf{F}}(\alpha : 1, v)]^{1/2}$$

Tables of the inverse survival function of the $\mathbf{T}$ variate are usually two-tailed so that the probability level shown is twice the α value.

2. For $v \geqslant 30$ the $\mathbf{T}$ variate $\mathbf{T}: v$ approximates to the unit normal variate $\mathbf{N}: 0, 1$.

$$\mathbf{T}: v \approx \mathbf{N}: 0, 1; \quad v \geqslant 30$$

3. Consider n normal variates $\mathbf{N}_i : \mu, \sigma;\ i = 1, \ldots, n$. Define variates $\bar{\mathrm{x}}$, s^2 as follows

$$\bar{\mathrm{x}} \sim (1/n) \sum_{i=1}^{n} \mathbf{N}_i : \mu, \sigma$$

$$\mathrm{s}^2 \sim (1/n) \sum_{i=1}^{n} [(\mathbf{N}_i : \mu, \sigma) - \bar{x}]^2$$

Then

$$\mathbf{T}: n-1 \sim (\bar{\mathrm{x}} - \mu)/[\mathrm{s}/(n-1)^{1/2}]$$

4. Consider a set of n_1 normal variates $\mathbf{N}_{1i} : \mu_1, \sigma;\ i = 1, \ldots, n_1$ and a set of n_2 normal variates $\mathbf{N}_{2j} : \mu_2, \sigma;\ j = 1, \ldots, n_2$. Define variates $\bar{\mathrm{x}}_1, \bar{\mathrm{x}}_2, \mathrm{s}_1^2, \mathrm{s}_2^2$ as follows

$$\bar{\mathrm{x}}_1 \sim (1/n_1) \sum_{i=1}^{n_1} \mathbf{N}_{1i} : \mu_1, \sigma$$

$$\bar{\mathrm{x}}_2 \sim (1/n_2) \sum_{j=1}^{n_2} \mathbf{N}_{2j} : \mu_2, \sigma$$

$$\mathrm{s}_1^2 \sim (1/n_1) \sum_{i=1}^{n_1} [(\mathbf{N}_{1i} : \mu_1, \sigma) - \bar{\mathrm{x}}_1]^2$$

$$\mathbf{s}_2^2 \sim (1/n_2) \sum_{j=1}^{n_2} [(\mathbf{N}_{2j} : \mu_2, \sigma) - \bar{\mathbf{x}}_2]^2$$

Then

$$\mathbf{T} : n_1 + n_2 - 2 \sim \frac{(\bar{\mathbf{x}}_1 - \bar{\mathbf{x}}_2) - (\mu_1 - \mu_2)}{\left(\dfrac{n_1 \mathbf{s}_1^2 + n_2 \mathbf{s}_2^2}{n_1 + n_2 - 2}\right)^{1/2} \left(\dfrac{1}{n_1} + \dfrac{1}{n_2}\right)^{1/2}}$$

Variate **W** : b, c
Range $0 \leqslant x \leqslant +\infty$
Scale parameter $b > 0$, b is sometimes referred to as the characteristic life
Shape parameter $c > 0$

Distribution function	$1 - \exp[-(x/b)^c]$
Probability density function	$(cx^{c-1}/b^c)\exp[-(x/b)^c]$
Inverse distribution function (of probability α)	$b\{\log[1/(1-\alpha)]\}^{1/c}$
Survival function	$\exp[-(x/b)^c]$
Inverse survival function (of probability α)	$b[\log(1/\alpha)]^{1/c}$
Hazard function	cx^{c-1}/b^c
Cumulative hazard function	$(x/b)^c$
r^{th} moment about the origin	$b^r\Gamma[(c+r)/c]$
Mean	$b\Gamma[(c+1)/c]$

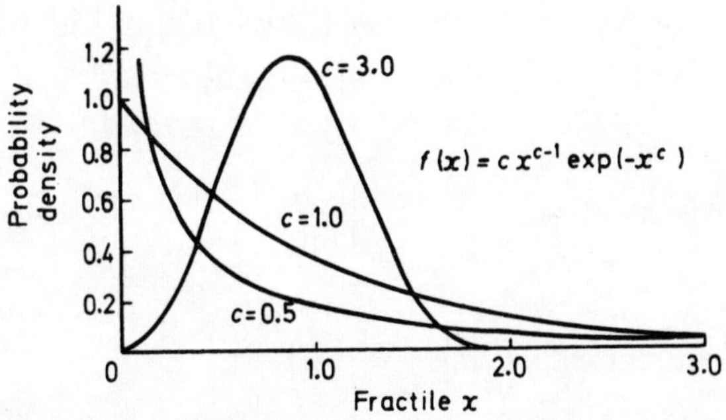

Figure 28.1 Probability density function for the Weibull variate W: *1, c*

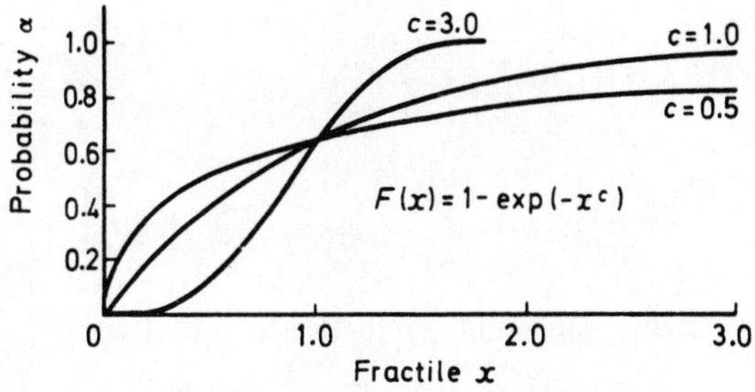

Figure 28.2 Distribution function for the Weibull variate W: *1, c*

Variance	$b^2(\Gamma[(c+2)/c] - \{\Gamma[(c+1)/c]\}^2)$
Mode	$b(1-1/c)^{1/c}, \; c \geqslant 1$ $0 \qquad , \; c \leqslant 1$
Coefficient of variation	$\left\{\dfrac{\Gamma[(c+2)/c]}{\{\Gamma[(c+1)/c]\}^2} - 1\right\}^{1/2}$

28.1 Parameter estimation

By the method of maximum likelihood the estimators, $\hat{b}, \hat{c}$, of the shape and scale parameters are the solution of the simultaneous equations

$$\hat{b} = \left[(1/n) \sum_{i=1}^{n} x_i^{\hat{c}}\right]^{1/\hat{c}}$$

$$\hat{c} = n \Big/ \left[(1/\hat{b})^{\hat{c}} \sum_{i=1}^{n} x_i^{\hat{c}} \log x_i - \sum_{i=1}^{n} \log x_i\right]$$

See also probability paper and cumulative hazard paper.

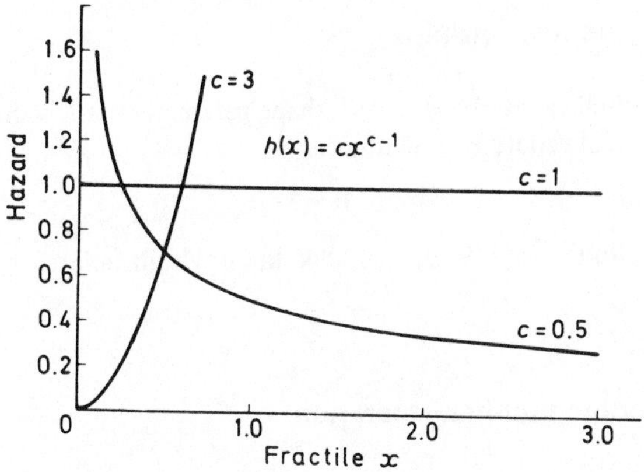

Figure 28.3 Hazard function for the Weibull variate W: *1, c*

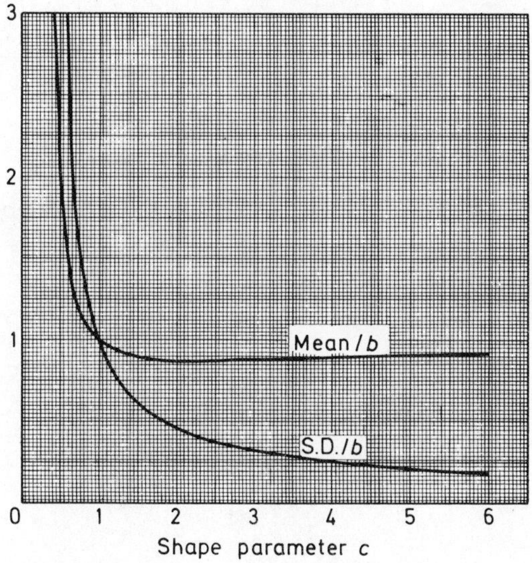

Figure 28.4 Weibull mean and standard deviation (S.D.) *as a function of the shape parameter c and scale parameter b*

28.2 Variate relationships

1. The Weibull variate **W** : b, c with shape parameter $c = 1$ is the exponential variate **E** : b with mean b.

$$\mathbf{W}: b, 1 \sim \mathbf{E}: b$$

2. The Weibull variate **W** : b, 2 is called the Rayleigh variate.

28.3 Random number generation

Random numbers of the Weibull variate **W** : b, c can be computed from random numbers of the unit rectangular variate **R** using the relationship

$$\mathbf{W}: b, c \sim b\,(-\log \mathbf{R})^{1/c}$$

28.4 Note

The characteristic life b has the property

$$\text{Prob}\,[\mathbf{W}: b, c \leqslant b] = 1 - \exp(-1) = 0.632 \text{ for every } c$$

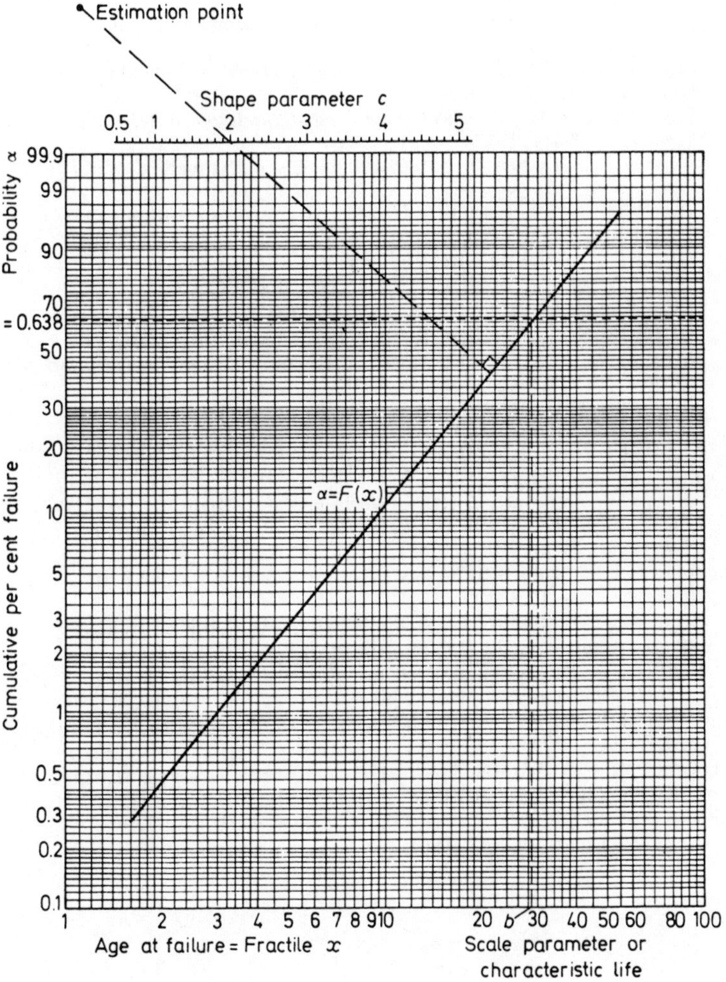

Figure 28.5 Weibull probability chart

BIBLIOGRAPHY

1. AITCHISON, J., and BROWN, J. A. C., *The Lognormal Distribution*, Cambridge University Press (1969)
2. BEYER, W. H. (Ed.), *Handbook of Tables for Probability and Statistics*, Chemical Rubber Co. (1968)
3. BROWNLEE, K. A., *Statistical Theory and Methodology in Science and Engineering*, Wiley (1965)
4. DUBEY, S. D., 'A New Derivation of the Logistic Distribution', *Naval Research Logistics Quarterly*, **16** No. 1, 37–40 (1969)
5. ELDERTON, W. P., and JOHNSON, N. L., *Systems of Frequency Curves*, Cambridge University Press (1969)
6. IRESON, W. G. (Ed.), *Reliability Handbook*, McGraw-Hill (1966)
7. JARDINE, A. K. S., *Maintenance Replacement and Reliability*, Pitman/Wiley (1973)
8. JOHNSON, N. L., and KOTZ, S., *Distributions in Statistics: Discrete Distributions; Continuous Univariate Distributions 1; Continuous Univariate Distributions 2,* Houghton Mifflin (1970)
9. KENDALL, M. G., and BUCKLAND, W. R., *A Dictionary of Statistical Terms*, Oliver and Boyd (1971)
10. KENDALL, M. G., and STUART, A., *The Advanced Theory of Statistics*, Vols. I and II, Griffin (1958)
11. KING, J. R., *Probability Charts for Decision Making*, Industrial Press (1971)
12. LLOYD, D. K., and LIPOW, M., *Reliability Management Methods and Mathematics*, Prentice-Hall (1962)
13. MYERS, B. L., and ENRICK, N. L., *Statistical Functions*, Kent State University Press (1970)
14. NAYLOR, T. H., BALINTFY, J. L., BURDICK, D. S., and CHU, K., *Computer Simulation Techniques*, Wiley (1966)
15. NELSON, W., 'Hazard Plotting for Incomplete Failure Data', *Journal of Quality Technology*, **1**, 1 (Jan. 1969)
16. OWEN, D. B., *Handbook of Statistical Tables*, Addison-Wesley and Pergamon (1962)
17. PEARSON, E. S., and HARTLEY, H. O., *Biometrika Tables for Statisticians*, Cambridge University Press (1956)
18. WEATHERBURN, C. E., *A First Course in Mathematical Statistics*, Cambridge University Press (1961)